SAFER OFFSHORE

Crisis Management and Emergency Repairs at Sea

by

Edward Mapes

Paradise Cay Publications, Inc.
Arcata, CA

Copyright ©2010 Edward Mapes

All rights reserved. No part of this book may be used or reproduced in any manner whatsoever without written permission, except in the case of brief quotations embodied in critical articles and reviews, For information, contact the publisher.

Cover design by Rob Johnson, www.johnsondesign.org
Editing, book design by Linda Morehouse, www.webuildbooks.com

Interior photographs by Edward Mapes unless otherwise attributed

Printed in the United States of America
First Edition
ISBN 978-0-939837-90-8

Published by Paradise Cay Publications, Inc.
P. O. Box 29
Arcata, CA 95518-0029
800-736-4509
707-822-9163 Fax
paracay@humboldt1.com

In dedication to The Water . . .

That avenue of such adventure that beckons mariners
to test resolve and mettle—threatened now as never before.
And to mankind . . .
in the hopes that we recognize the peril in time.

Table of Contents

Acknowledgments — v
Introduction — vi

1. Communications — 1
2. Running Aground — 13
3. Towing — 25
4. The Iron Jenny — 41
5. Rigging Failures — 61
6. Loss of Steering Control — 99
7. Sail Repair — 125
8. When the Boat Floods — 135
9. Fire at Sea — 147
10. Crew Overboard — 159
11. Abandon Ship! — 173
12. Helicopter Evacuation — 213
13. Restricted Visibility — 221
14. Heavy Weather at Sea — 231

Appendix
    Tool Kit and Spares — 275
    Monitoring and Maintenance Checklist — 280
    Voyager's Deck Log — 281
    Contributors' Contact Info — 282
    Beaufort Scale of Wind and Sea States — 283
Index — 284
About the Author — 292

Acknowledgments

The author would like to recognize and thank the following people and organizations: Travis Blain, of Mack Sails; the United States Coast Guard; Landfall Navigation; Speed Plastics Limited; Tom Rau, author of Boat Smart Chronicles; Blue Water Marine Engines; Mike Meer, of Southbound Cruising Services, LLC; Rig-Rite, Inc.; National Oceanic and Atmospheric Administration; Zack Smith, of ParaAnchors by Fiorentino; Jess Gregory, of Banner Bay Marine, LLC; and the National Data Buoy Center.

Their generous contribution of expertise, advise, suggestions, and photographs have proved to be instrumental in the production of this book. Their contact information can be found in the Appendix.

Introduction

You could say this book has been thirty years in the making—that's how long I've been taking boats to sea. The inevitable truth became apparent on the very first passage: things break on boats. There's often no rhyme or reason, but it will happen just as sure as the tide flows out. Seagoing captains adopt methods that help them cope with the nature of sailing offshore: they equip, inspect, and prepare their vessels for voyaging, and they adopt attitudes of independent self-sufficiency. They use experience, common sense, and confidence to manage problems as they inevitably arise.

Ever wonder why the best sailors are the ones who seem to have the fewest mishaps while on the water, the fewest horror stories to relate? Often it's because they've detected and corrected defects on their vessels and gear before shipping out. Their crews know the boats and don't make mistakes that less-seasoned crews make. The vessels are monitored well while under way, and irregularities are spotted early before they can multiply into larger problems. And these fine sailors tackle incidents with a seamanlike attitude of calm and confidence, knowing that it's all part of sailing the seas. Their logs might not even make note of a situation that would have been a major incident for less-seasoned sailors.

The misfortunes of many sailors and their vessels, sometimes with the very worst consequences, are often the result of a cascading series of events that lead to dangerous circumstances. Sometimes the real art of competence is in recognizing situations as they occur, and having the good sense and ability to interrupt the chain of events before it leads to real trouble.

"Jury rigging" is a term handed down through sailing lore. Try to imagine situations that arose on the wooden sailing ships of the day. Those vessels didn't enjoy the construction methods, materials, and technology with which we are blessed today, but they were sailed by true masters who understood that a ship at sea was a world unto itself. With no satellite phone or EPIRB to summon help when the ship or crew were in trouble, those masters repaired their ships, navigated, and managed crew injury/illnesses with the resources available on the boat. They became adept at recognizing a problem, attending to it immediately with all the resources they could muster, and preventing it from escalating into a serious threat. That small tear in the mainsail foot was attended to long before it ripped the sail in two as the first gale winds took hold of it.

We're fortunate to sail vessels so advanced, fitted with technological wonders in communications and navigation that our predecessors could not have even dreamed of, but our reliance on those marvelous systems can erode basic sailing skills and the spirit of independence that marks the best passage-making seamen.

This book discusses the art of the jury rig, but the best fix is one that never takes place because our foresight prevents its necessity. The truth is, I've seen breakdowns of virtually every system and piece of gear you can name during my time on the water, and evidently devised ways of surmounting those problems well enough to be sitting at this computer. I wonder how many breakdowns could have been avoided had I been this vigilant in those early years of offshore deliveries. Sailing the oceans and coming to understand the importance of preparation was the inspiration for my first book, *Ready to Sail*. Now I, along with many other captains, have adopted this methodical approach to make sure that every boat we sail is prepared in advance. We make sure that she's been thoroughly inspected, all defects have been attended to, essential gear has been added, and the crew has been thoroughly oriented before we leave port. Experience has shown, over and over again, that breakage, injuries, and mistakes happen less when the boat and crew are truly "ready to sail."

Forethought, preparation, constant monitoring while under way, experience and the confidence it brings: these are the elements of a sailor. This book is written as a reference to use when the inevitable breakdown occurs to those far away from help. We'll discuss common problems that arise and ways to fix them. But keep in mind the more important art of preventing accidents and breakage from happening in the first place, and assuming the attitude that the responsibility of maintaining our vessels and crew lies not with whomever we can reach on the SSB; it lies with us.

When I think of the term jury rig, I picture a boat with only half of its mast. It's limping into port; a block is secured by lashings to the jagged upper remnants of the spar. A halyard through the new sheave hoists a jib's tack aloft, up the abbreviated spar. The sail's head is made fast at the stem, and a sheet tensions the clew so that the sail captures some wind and the boat makes slow way toward home.

We're going to expand the meaning of jury rig in this book. Along with describing methods to fashion a makeshift sail on a broken spar, we'll discuss a variety of repairs and rigs of all kinds. Some of them aren't as dramatic or sexy as limping into port with a broken stick, but they'll help you make landfall nonetheless.

In my description of true sailors, I mentioned that they use forethought in readying themselves for upcoming passages. When I first began learning crew overboard tactics—predominately the Figure of Eight technique in those days—I made a habit of thinking about the steps to take each time I took the helm of my own boat or one I was racing. It always felt better knowing that I could respond without hesitation should the situation arise. After using this technique of study, the steps to take for victim recovery on each point of sail began to come back to me automatically, as mind memory—as opposed to muscle memory—was established. Soon I didn't have to take the time to consider the options at all; it had become second nature.

Taking stock of our boats and considering what could go wrong is a similar exercise that helps us when responding to trouble on the water. What would I do if the rudder broke? How would I cope if a shroud terminal cracked and was about to give? If the autopilot failed, how would I go about fixing it?

Asking these questions in advance is a great way to make ready for sea. We might not prefabricate replacements for everything that could break, but the exercise provides us the chance to have materials and tools ready to build a rudder, for example. We could have a plan in mind to substitute for a shroud and keep the rig standing. We would know what goes wrong with a wind vane, what to look for, and how to return it to use.

You don't usually notice or even hear about the many vessels that had their share of problems offshore. The repairs were completed and they made landfall quietly and efficiently. These sailors most likely had the skills, materials, and a plan to cope with such situations, and merely had to carry out the work to get back on course.

Very few boats have the room to house a full workshop, with all the tools and materials we'd like to carry. Our tool kits must contain the elements that we usually use, and items to do some specialized jobs that could be called for. For example, if removing an impeller from the raw water intake pump requires a certain type of wrench, that wrench needs to be on board. If it takes a tiny screwdriver to repair someone's reading glasses, it should be in the kit. Should the engine have 5/16" fuel line from the tank to the primary fuel filter, there should be a length of 5/16" fuel hose available to replace it. This is where exercising forethought helps when it comes time for "McGuivering"; it takes less ingenuity when the right tools/materials/skills are available.

Just as we can't have all the tools and materials on a boat that we might have in our workshop at home, neither can we foresee every breakdown that can occur at sea. What is more, going over our vessels before departure cannot obviate every eventuality once we make the open ocean. In truth, Murphy's Law governs on the water just as it seems to on land; the sailor best able to accept that reality and deal with its consequences without elevated stress levels will enjoy life at sea the most.

Even the most vigilant predeparture inspections can't reveal flaws that haven't yet arisen. The possibility that we'll have to fly (sail) by the seat of our pants and employ some ingenuity always exists. The ocean has a way of finding our weak links, and sometimes likes to challenge us. So a spinnaker becomes snarled and wedged into a spreader/shroud intersection when the wind catches it during a hoist. It won't go up, nor will it come down. Neither had we planned specifically for the autopilot sensor arm to disconnect itself from the steering quadrant after 18 hours of rough weather, but it did just the same and now you've got to deal with it. These are examples of situations that come up out of nowhere when you were just trying to have a nice day, but aren't they part of the reason we take our vessels and crews on these journeys? Doesn't that have something to do with the lure of testing ourselves on offshore passages? Isn't that part of the great sense of achievement we feel when making landfall? Of course it is, and those that are too intimidated to try it will never experience that gratification.

The approach to problem solving is to quickly assess the situation without allowing the adrenalin to turn your brain into Sargasso weed. With the snarled spinnaker, place the boat on the most stable tack for the conditions, control speed, and send an able person safely aloft to untangle the spinnaker. Then bring it back to the deck and resume course.

Refer to the Appendix for a tool kit and lists of spare parts and materials that you might want to carry on extended passages. Your consideration of what could go wrong will show you other tools and materials that may not be listed, and your level of readiness will improve.

When I inspect a boat during the preparation phase of an offshore project, I concentrate on each system individually, making notes in a checklist as the inspection proceeds, and then move on to the next system. This workmanlike method helps keep the overlooked defects to a minimum. In the upcoming chapters we'll discuss breakdowns and their fixes in a similar fashion, pointing out likely failures of individual systems and how to make repairs.

As you consider gear and equipment additions, bear in mind this simple truism: the more gadgets, toys, and stuff you bring on board, the more breakage and failures you'll have to worry about and fix. Electrical appliances in the galley, electronic charts, electric winches, furling main sails, multiple refrigerator/freezers, etc. are all nice conveniences, but they eat electricity and are prone to failure in the marine environment. I suggest that you factor that into your formula for gearing up the boat.

1

Communications

I sailed for a lot of years without long range communications equipment, but won't be in that situation again, for I consider it mandatory for anyone sailing great distances from shore. When I single-handed *Voyager* from Tortola to Annapolis in June–July of 2003 without contacting home, those 10 days offshore proved to be more difficult for my family than they were for me. That voyage demonstrated to me that regular communication with the home base is not only prudent regarding safety, but it also alleviates the worry and tension that loved ones feel most acutely.

I've been aboard vessels with single sideband radios and satellite phones, and have opted for satellite communications for a number of reasons. It's important to me that were *Voyager* ever sunk or dismasted, I would still have long range communications. SSB would be lost because it uses the vessel's backstay or copper strips within the hull for its antenna. Another important consideration is that the satellite phone (with a portable antenna) would go with us into the life raft if that ever became necessary. I have programmed emergency numbers into the unit and have a waterproof list of contacts in the overboard bag as well, so it would be easy to contact home, the Coast Guard, emergency medical assistance, or Mom from adrift. Along with a handheld GPS, communicating our situation and position to rescue agencies should be assured.

Combined with a laptop computer and communications software, the satellite phone can be the conduit to e-mail messages and weather information, the two most important components of offshore communications. Satellite connections and easy communication have never been a problem with either Iridium or Inmarsat phones that I've used. I highly recommend installing the external antenna, though: it provides a more reliable connection, and is much easier and more convenient than holding the phone upright, often on deck, while waiting for contact to be established.

The weather information and e-mails arrive instantly after your request is received, and they are easily viewed on a laptop. From there, the information can either be saved or printed out for easy viewing and discussion with the crew.

Contacting other parties—the family at home, a doctor, the Coast Guard, or your mother—can be done either via e-mail or by direct dialing their land-based telephones. E-mailing is done just like from a computer at home; just insert the address and type your message, send it to the software, and it's delivered when you make the next satellite connection. Messages should be batched together to save on satellite air time.

Conversations are crystal clear when the phone is used to dial another telephone. It always amazes me that the voice sounds as though I were listening to my neighbor from home. I can tell you that being called by someone from the middle of an ocean means a lot to folks back home.

Figure 1.1 (opposite) illustrates *Voyager*'s stern post, which accommodates the satellite phone's external antenna. This location situates the antenna with a clear view of the sky, yet close enough for easy access to service.

Offshore sailing vessels should be equipped with both a hard-wired VHF radio and a battery-powered handheld unit. The primary ship's VHF is capable of transmitting at either 1 watt or 25 watts of power, while handheld units send at 1 or 5 watts. Use the 1 watt setting for contacts in high-traffic areas so that your call is heard

Fig. 1.1
The stern platform provides an ideal platform for the radar, wind instrument, and satellite phone.

by fewer eavesdroppers and to minimize use of the airwaves. The 25 watt setting is necessary for more distant communications, up to about 25 miles. Either radio is useful for communication inland and on the coasts, but the more powerful ship's VHF is usually necessary for ship-to-ship contacts offshore.

The handheld is still very important on an offshore passage; it is a key element of the Abandon Ship bag that would be taken into the life raft should it become necessary to leave a distressed vessel. It's wise to have a portable VHF antenna on board should the masthead unit be lost.

Computer

Computers have increasingly become standard equipment on passagemaking and cruising boats. Where to keep the computer is a key decision. It should be accessible, but must be kept dry and safe from jostling. Computer failure is usually due to being dropped or banged about, rather than from corrosion. A splash of seawater through an open port, though, will usually spell doom for the unit. RAM Mounts from National Products, Inc. (www.rammount.com) is a source of convenient mounting solutions.

Good ventilation is a major consideration, given the high humidity and salt content of the atmosphere. Add ventilation if the computer is stored in a locker. A laptop can be powered by a 12-volt adapter or a small inverter dedicated to the computer to assure a clean power source. You should be able to charge the battery with either DC or AC power on board. A small, uninterruptible power supply (UPS) will maintain power to a desktop (or other AC-powered systems) during voltage-drop periods and allow for orderly shutdowns if the inverter fails.

For those intending extended cruising, make sure to take along CDs that came with the computer, especially ***the manufacturer's system recovery CD.*** Bring all charts for electronic navigation to reload as needed. You'll also want your serial code or security codes to install them.

Make a list of all technical support phone numbers you may need. Bring along any written technical or software support information, as well as installation or usages guides, or print such information out.

Don't forget to include extra cables, adapters, a long phone cable and computer network cable for getting connected in places where the data outlet and power plug are at opposite ends of a room. Carry extension serial or USB cables and a spare battery for the laptop, along with AC voltage and plug adapters.

Communications

Your goal when using the computer at sea is to enhance efficiency and decrease online time. Make downloading information as simple and fast as possible. Adjust computer settings beforehand; don't waste time doing it at sea.

Consult a computer technician for assistance in turning off the routine updates to computer software and services. These include such things as automatic stock or weather reports, or new software versions. Turn off any service that moves data over the internet without your asking for it.

Turn off the Windows Error Reporting function. Download the latest security patches and anti-virus updates, then disable all Windows or Apple Automatic Updates. Re-enable these functions after the sea voyage.

You can compensate for difficulty in viewing the LCD display screen by making the background pure white. This enhances the contrast between colorful icons and the white background.

Once the computer is working well with whatever devices you choose, such as GPS or a satellite phone, use the Windows System Restore feature to place that configuration in memory. If the computer setup changes or crashes, this utility will allow you to retrieve the desired configuration.

When medical emergencies, heavy weather, or the need to abandon ship threaten a vessel and its crew, expeditious communications can be crucial to survival. Advanced technologies in marine instrumentation have increased the efficacy of marine communications when we need those most: during critical situations.

Marine communications can be subdivided into two broad categories: those from inland lakes or within 25 miles of the coast, and contacts initiated from vessels offshore, out of VHF range.

Rescue 21 is a Coast Guard initiative to enhance contacts with vessels in distress up to 25 miles from the coasts and

5

inland. This is a nationwide radio communications network that takes advantage of new technologies and equipment to make distress calling faster and more efficient than ever. The Coast Guard network has been equipped with updated marine radios, computers, and antennas; to take advantage of Rescue 21, boaters must also upgrade their communications equipment.

DSC, or Digital Selective Calling, represents a global upgrade in maritime distress communications, and is at the heart of the Rescue 21 system. DSC/VHF radios utilize the latest in satellite and digital technology previously seen only on commercial shipping but now available to recreational boaters. DSC/VHF radios allow for direct ship-to-ship private communications between vessels, but the major advantage is in sending an automatic distress signal in times of emergency. The DSC radio operates much like an EPIRB that sends encoded Maydays directly to satellites.

When the Distress button is depressed for 5 seconds, the DSC radio emits a powerful coded digital signal that punches through adverse conditions within 1/3 second. The signal, transmitted on channel 70, occupies its own frequency, and eliminates interference from other transmissions. Once transmitted, the signal:

- Identifies your vessel.
- Identifies you.
- Provides latitude/longitude position of your vessel (when connected by a 2-wire bridge to your GPS).
- Sounds a distinctive alarm at all area Coast Guard facilities and on all other vessels with DSC radios.
- Automatically switches your radio to channel 16.

The Coast Guard responds to the distress signal, and initiates voice communications with your vessel on channel 16. The DSC radio will also continue sending the emergency signal if the skipper is disabled.

Your unique Maritime Mobile Service Identity, or MMSI*, must be entered into the radio to enable ship identification during distress messages. The radios are also programmed with a unique nine-digit FCC identification number that allows the ship-to-ship calling feature. Registration of the radio with the FCC enters that information and your boat MMSI in the US Coast Guard's national distress database.

Current VHF marine radios without DSC still allow boaters to contact the Coast Guard on Channel 16. They monitor Channel 16 on a 24-hours-a-day, 7-days-a-week basis. This hail will be heard by the nearest Coast Guard Station or vessel.

When emergencies arise on vessels operating more than 25 miles offshore, the DSC/VHF radios will be out of range, except for other vessels in the vicinity. Communications there must utilize either satellite telephones or Single Sideband Radio technologies. Several Single Sideband frequencies are designated for use in distress situations.

SSB Distress Frequencies

Band	SSB Frequency
MF	2182 kHz
HF4	4125 kHz
HF6	6215 kHz
HF8	8291 kHz
HF12	12290 kHz
HF16	16420 kHz
VHF	156.800 MHz

Marine SSB Distress Frequencies

Modern Single Sideband Radio transceivers are also combined with Digital Selective Calling to enhance offshore distress and ship-to-ship contacts. This combination of technologies operates in a fashion very similar to DSC/VHF, but with a much greater range. During an emergency, the radio's red Distress button is depressed for five seconds, initiating the automated call for assistance. The ship's identification and positioning information, nature of emergency, and time of transmission is broadcast to coast stations, large ships, and commercial vessels at sea. The transmitter operates on all of the marine bands from 1.6 to 27.5 MHz at 150, 60, or 20 watts peak envelope power. The radio can transmit on any one of the six DSC distress frequencies or in sequence on all of them, and the call automatically repeats at intervals of 3-1/2 to 4-1/2 minutes until another vessel answers or the vessel in distress cancels.

Band	SSB/DSC Frequency
MF	2187.5 kHz
HF4	4207.5 kHz
HF6	6312 kHz
HF8	8414.5 kHz
HF12	12577 kHz
HF16	16804.5 kHz
VHF	156.525 MHz (Ch 70)

Marine SSB/DSC Distress Frequencies

The Coast Guard is still my first choice for contact during emergencies, even when we are hundreds of miles from any land. The Coast Guard Atlantic and Pacific Area Command Centers are subdivided into District Command Centers (DCC), each with specific regions of responsibility. Vessels equipped with satellite phones can place calls directly to the Coast Guard at a variety of locations to initiate distress responses. Mariners in distress placing calls to the Coast Guard for emergency assistance can contact the Coast Guard Marine Safety Center, Atlantic or Pacific Regional Command Center, or the nearest District Command Center.

U.S. Coast Guard Contact Numbers

   Norfolk Command 757-398-6390
   I have this number programmed into my satellite phone for easy use if we ever have to abandon ship. This command center can field your call and initiate communications to any appropriate agency.

   Marine Safety Center
   202-475-3400
   202-475-3403
   202-267-2100 (Monitored 24/7)

   Central Command: 800-DAD-SAFE

   Atlantic Area Command Center
   (Great Lakes, Gulf, and East Coasts): 757-398-6231

   Pacific Area Command Center
   (Hawaiian, Alaskan, and Pacific Coasts): 510-437-3701

District Command Centers around the country, along with their emergency contact numbers:

Emergency DCC Contact

Designation	Location	Number
1	Boston, MA	617-223-8443
5	Portsmouth, VA	757-398-6376
7	Miami, FL	800-874-7561
7	San Juan, PR	787-729-6800
8	New Orleans, LA	504-589-6225
9	Cleveland, OH	216-902-6283
11	Alameda, GA	510-437-3701
13	Seattle, WA	206-220-7001
14	Honolulu, HI	800-842-2600
14	Guam	671-355-4910
17	Juneau, AK	907-321-4501
		907-321-4503
		907-321-4510
		907-321-0247

It is important to have the capability to charge electronics such as the satellite phone with either DC or AC power (land-based or through the inverter). This gives assurance that if one mode fails, charging is still possible. It is possible to jury rig power into a phone, however, which I had to do several years ago. We had lost electrical battery power, and the phone battery was quite low. Though the experience taught me that you never allow the phone battery to discharge, I was able to connect a battery from the handheld VHF (Fig. 1.2, following page), enabling us to send several e-mails and download some weather charts four days out from making landfall.

Loss of an antenna can cripple any radio. VHF and satellite phones connect to their external antennas with coaxial cables. There are several portable antennas available at Radio Shack that can be used as an emergency replacement for either of these phones.

COMMUNICATIONS

Replacing a Single Sideband Radio antenna is a bit trickier, but could be extremely important should the rig be lost. To rig a replacement antenna, use several feet of 14 gauge copper wire that connects to the SSB tuner. The wire must be rigged vertically, so you'll need a mechanism to hoist and keep it in position. That could be the mast, a portion of the mast, or a post such as a spinnaker or whisker pole used to hold the wire erect.

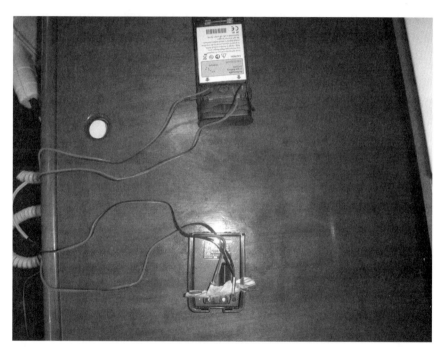

Fig. 1.2
Here, we used a handheld VHF battery to power the satellite phone after its DC charger malfunctioned.

More modern technology has improved our communications with land bases and rescue agencies, facilitating position tracking by anyone with internet access. Consider this: others will be able to watch your progress, in real time, across an ocean by simply logging onto the net and putting a link into the browser. What is more, the device can easily be transferred to a life raft in the event that you have to abandon ship; your position information would be broadcast just the same, making rescue a matter of traveling to your coordinates. This becomes possible with a satellite tracker/messenger, now commercially available for purchase or rental from several companies.

These devices enable the transmission of text messages to cell phones or e-mails to friends and family members and can also send your coordinates and a distress message to appropriate rescue agencies all over the world. In my view, this revolutionizes the prospect of life raft survival.

*Obtain your MMSI number for free at www.boatus.com/mmsi/instruct.htm

2

Running Aground

There are two types of sailors: those who admit to having run aground and those who have grounded but lie about it. Embarrassing? Yes, it can be, but we all touch our keels to the bottom sometimes. What distinguishes the best grounders, however, is whether or not they know how to free themselves. I'd like to share what I've learned about being "under way but not making way"— yes, some of it from personal experience.

Touching bottom can be semi or totally intentional: for example, when proceeding very slowly in shallow waters in search of an anchorage or when running aground in the event of a hull breech. For the most part, though, groundings are not scheduled, yet there are many precautions and measures we can take to prevent them.

Obviously, our chances of remaining afloat improve when we know about the water depths and protruding objects where we sail. Harbor charts are invaluable sources of information about water depths, tidal ranges, bottom characteristics, positions of shoals and bars, shallow areas, etc., and should be consulted whenever nearing an unfamiliar anchorage.

Unless absolutely necessary, never enter an unfamiliar port after dark or during poor visibility. Groundings occur most often during these unwise approaches. It makes a lot more sense to either delay your arrival by decreasing speed en route or to heave-to a few miles from the harbor area until daylight. Whenever entering a harbor or channel, honor all navigational buoys (if they exist), decrease speed so that rapid stopping is possible, watch the depth gauge on a constant basis during the approach, and post a bottom watch to observe changes in bottom color and contour. These measures should all be standard operating procedure. I have climbed to the spreaders more than a few times to watch the bottom when approaching islands.

Stay away from prominences of land such as spits and bends in waterways; the innermost bend is often shallow water. I once grounded a heavy Vagabond 42 on a shoal area adjacent to such a curve in the Ashtabula River, off of Lake Erie. This boat was a rarity in the Great Lakes, and a small crowd gathered on shore as the Coast Guard boarded us, did their inspection, and then pulled us off, red faces and all.

Tide charts and tidal current charts, the Coast Pilots, cruising guides, etc. are important sources of information when entering coastal harbors, especially those with large tide ranges, currents, or troublesome entrances. You'd like to make your approach during slack high tide if possible so that the depth is maximum and current is minimal. This is certainly the wrong time to go aground, though, because when the tide goes out, you'll really be stranded. The approach to some rivers should only be attempted on the flood or slack stages, because an ebb tidal current flowing against predominately adverse winds sets up dangerous conditions.

I once had a wild ride just before nightfall while attempting landfall in Beaufort, North Carolina. The Coast Pilot specifically warns about the Beaufort inlet, and on this occasion I thought I'd beat the ebb. Crossing the Gulf Stream drove me farther to the north than planned, though, and it took longer getting to the outer channel marker than I'd anticipated. A 20+ knot breeze from the east opposed the outflowing tide and created short, steep waves that seemed to swirl about. After reaching only the first buoy, with the boat heeling and yawing wildly, I spun around, sailed out a few miles, and hove-to until the ebb had stopped its flow.

I once observed an unfortunate vessel that had gone aground during the night entering the harbor at Port St. Napoleon, France at high tide. By the next morning, as we departed port to traverse the Mediterranean, we saw two harbor patrol boats along with a salvage vessel on scene, all straining to free this 40-foot wooden sailboat that was high and dry. I expected to see the planking start popping off the hull, as it was surely distorted from the pull of those boats.

The first action to take upon becoming grounded is to ensure crew safety; injuries can easily occur when the front of the keel suddenly stops and crewmembers are sent flying.

Once you've looked into the well-being of the crew, immediately inspect all bilge areas and keel bolts for water leaks. Boat damage ranges from gaping holes in the hull to more subtle defects to the hull/keel junction. Contact with the bottom can force the aft section of the keel upward against the hull. The keel/hull junction can be disrupted, and water integrity lost (Fig. 2.1, following page). Efforts to extract the boat can result in lateral forces that wrench the keel from its secure hull attachment. The impact can also crack gelcoat and fiberglass extending out from the keel (Fig. 2.2, following page).

Keel bolts can be dislodged or can loosen as the shock transfers through the keel. You will then see them depressed into the hull, or protruding higher than normal. Listen when you're under way; a dislodged keel can make audible noises during maneuvers.

The Convention on the International Regulations for Preventing Collisions at sea of July 15, 1977 (COLREGS) governs

Fig. 2.1
This grounding impact drove the aft keel up into the hull. Major repairs were necessary.

Fig. 2.2
Gelcoat cracks after grounding.

the day markers, lighted buoys, and sound signals displayed by all vessels. If the grounded boat is less than 100 meters in length and poses a risk to other vessels, or if the other vessel should be warned of your situation, be prepared to display the following:

—In unrestricted visibility, a vessel aground shall display three balls in a vertical row.
—In restricted visibility, a vessel aground shall, at intervals of not more than one minute, ring a bell rapidly for about 5 seconds. In addition, give three separate and distinct strokes on the bell immediately before and after the rapid ringing of the bell. A vessel aground may in addition sound an appropriate whistle signal of one short blast, one prolonged blast, and one short blast.

The majority of groundings are minor in nature, without serious damage or crew injuries. The crews are often able to refloat their boats using the methods described below. Sometimes, however, extenuating circumstances such as an ebbing tide, wave action, or holing of the vessel make grounding a dangerous situation (Fig. 2.3, page 18).

If a grounded vessel is pushed inshore by waves from seaward, set anchors to prevent being pushed farther toward the beach and to prevent the boat's being broached. Inspect the hull from outside if possible and survey for damages. Once the boat is stabilized, make a careful assessment of the situation and contact rescue agencies or the Coast Guard for assistance.

Grounding in this manner during storm conditions is especially dangerous for the crew and the vessel (Fig. 2.4, page 19). Use extreme caution when sailing in rough weather near a lee shore or making a harbor approach.

Fig. 2.3
***This barge has been driven aground in severe conditions,
with powerful wave action that puts the vessel in severe jeopardy.***

The initial grounding impact can be severe, with great risk of crew injury. The hull is also subject to more damage upon impact. The keel and rudder are most susceptible, since they extend farthest into the water. The amount of damage also depends on the type of bottom upon which the grounding occurs—sand, mud, coral, rocks, an island, and so on. Sand and mud are most forgiving; coral, rocks, and islands are the least.

Incoming waves can pound against the hull, driving the boat farther onto the rocks, worsening the damage or leading to loss of the boat (Fig. 2.5, page 19).

Make immediate attempts to get off and into safe water. Heel the boat, use the engine, trim sails, use lines to an anchor, get help: whatever is necessary and possible. The wave action may actually be helpful in lifting the boat, if you can get it to move while lifted off the bottom.

Fig. 2.4
This sailboat became grounded on an ebb tide.
Notice the list that has developed as the tide flows out.
Water is nearly over the lee decks.

Fig. 2.5
This vessel was driven onto shore by wave action
after she initially ran aground.

If the hull is holed and taking on water, running her intentionally toward a safer beaching area may be the best solution. Use all means at your disposal to get the boat free of coral or rocks. Steer her toward a sandy area and run her aground.

If waves lift the boat and drop her repeatedly onto the hard bottom, more damage is certain to occur. In this dire situation, the only way to prevent more hull damage may be to scuttle the boat, sinking her to the bottom. Set anchors out to deeper water, and point the bow out if possible. Make sure all crewmembers are able to get off the boat and safely to shore. If time permits, remove any gear or belongings that will be ruined by water, and then open seacocks and remove their connecting hoses. The boat will flood and sink to the bottom. This is a last ditch effort to salvage whatever is possible. The engine and woodwork may be ruined, but the hull, rigging, sails, and gear may be saved. If the tide ebbs and storm abates, the water may drain from the boat and you can assess the situation further at that point.

If the grounding does not occur on a lee shore, check for crew injuries and then inspect the bilges and keel bolts. If the tide is flooding, it may lift the keel off the bottom and allow an easy escape. The tide or an adverse wind may also carry the boat farther into the shallow zone and make the situation worse. If the boat is drifting toward shallower water, immediately set an anchor to control the situation and consider options to take advantage of the rising waters.

There is always the urge to start the engine and gun it at high RPMs to power the boat off in reverse gear. This may be successful, but it must be done carefully. The boat should not be backed for a prolonged time because of the risks of taking fouled water into the engine's cooling system and washing more sand under the hull. If the boat does not make sternway after a few seconds at high throttle, discontinue the backing and consider alternative methods.

If the wind is at an advantageous angle, hoist the main sail and trim it in. The weather helm may spin the boat toward deeper water. If so, hoist the jib and trim it for speed. As the keel lifts with a rising tide, the boat should gradually make her way to deeper water. Move all crew weight to the low side to heel the boat as much as possible, decreasing draft as she heels.

It may be possible to reach another vessel for assistance; be careful to make an agreement that the situation does not meet the conditions to qualify as salvage. Also, find out if the other captain intends to charge for the assistance, and if so, agree on a fee in advance.

The other boat may be able to tow you off with a line that you provide from a stout attachment at the bow, unless another location is more suited. Heel the boat over with crew weight, place the rudder on the appropriate heading to deeper water, and allow the towing vessel to make its attempt. See Fig. 2.6, below. This schooner is being assisted by two Coast Guard vessels. The halyards are being hauled in an effort to increase heel to break free of the bottom.

Fig. 2.6
This schooner is being assisted by two Coast Guard vessels.

I'm always concerned with the laterally-directed pull on the mast during this operation, and advise that a second halyard taken to the opposite toerail lends support against the strain taken on the active halyard.

Without a second vessel to lend assistance, you can use a dinghy in an attempt to tow the boat, or to take a halyard outboard to heel the boat far enough to bring the keel off of the bottom. Once the vessel floats, use the engine to carefully move it toward deeper water.

The dinghy can also be used to take an anchor toward deeper water or to the windward beam. Place the anchor in the dinghy carefully so that the weight doesn't turn it over, and be careful not to puncture the dinghy. Take the anchor out to the maximum extent of the rode, and then carefully drop it while counterbalancing the dinghy to prevent it from capsizing. Once the anchor hits bottom, the deck crew then hauls in on the rode to set the anchor with a primary cockpit winch. After the anchor sets and the line gets taut, continue hauling to kedge the boat off. The anchor can also be tied to a halyard from the mast head. When the halyard is winched in, the anchor will pull the mast down to heel the boat and swing the keel up and off of the bottom.

If you've grounded during an ebb tide, actions will have to be very rapid to get the boat free before the outgoing tide leaves you high and dry, waiting for the next flood (Fig. 2.7, facing page).

If a burst of the engine fails to free the boat, even when all the crew weight and sail power is used to maximize heel, have some crewpersons immediately get the anchor into the dinghy and haul it out to deeper water. Either kedge the boat toward deeper water, or tie the anchor line to a halyard to aid in heeling the boat. Keep the boat heeled to decrease draft, and occasionally gun the engine while observing the boat's position relative to a fixed object on shore. Don't hesitate and don't stop until the boat is safely afloat.

GROUNDING

Fig. 2.7
***This boat took out the day marker and grounded hard
on solid rocks. The ebbing tide left her stranded.***

If the tide ebbs too quickly or you're grounded too securely, then the best you can do until the next flood is to prepare the boat for a stay on the bottom. If the bottom is sand, mud, or weeds, your main concern is to face the mast uphill, away from deeper water. If the boat leans downward, she might tilt very steeply as the tide ebbs, and waves or a rising tide could force water over the gunwales.

If it's impossible to maneuver the mast toward the shallow, all seacocks should be closed, and house batteries switched off. Check the locations of tank vents; they may take on water when the levels rise. If the boat lists far enough, sections of cockpit could even flood with rising water.

Make certain that you know the exact draft of the boat. Depth sounds can be calibrated to show depth below the keel or below the hull, and that difference could become very important. If you're not sure of the settings, take advantage of the shallow water and extend

23

a line, boat hook, or measuring ruler to the bottom of the keel for a precise determination of the keel depth. This will be useful once the tide reverses.

While the boat lies high and dry, be sure to place an anchor to prevent the rising water from driving her farther aground. As the water fills, the boat will gradually rise up off of the bottom and into an upright position. Use your depth sounder or a lead line to measure the depth. Once the depth is close to your draft, heel the boat with crew or a kedge line to an anchor, hoist the sails, and start the engine. Have a crewmember check the raw water intake filter for sand or debris. Shut the engine off until the depth increases to a point that the intake port is clear, to aspirate only clear water. With the engine running, sails trimmed, and boat heeled, watch carefully for the water levels to elevate the boat off of the bottom. Once she starts moving, observe closely, and use the engine sparingly if necessary to aid in progress. The boat should eventually begin to make way and eventually reach safe water.

Other measures you can take include removing gear and personal articles and emptying water tanks to lighten the load. This could especially be necessary if winds prevent the normal tide heights or you are grounded during high tide. You may also require towing assistance in addition to flood tides if the boat sinks into very soft mud and must be broken loose from its stubborn hold.

3

Towing

The bulk of my towing experience has been while sailing the Etchell's class racing boats. These are pure one-design boats that sail fast but don't have engines. When wind conditions are light, they're often towed out to the racecourse, sometimes several boats in a row, by motorboats. Towing under these circumstances was simply a matter of securing a line from the towing boat around our mast and letting the other guy do the work. Towing can be very difficult and dangerous under other circumstances, though, and must be carried out properly.

Towing a vessel in crowded waterways, with current or tide running, during storm conditions, or in limited visibility, can make a bad situation worse when things go wrong. If you don't have the right gear and/or experience for towing, it's probably better to stand by the stricken vessel, radio for help, and wait for more qualified assistance. In remote waters, though, this might not be an option.

The two boats must establish radio communication before towing operations begin. It's important to understand the nature of the disabled boat's problems, any damages it's sustained, whether it can maneuver with its engine and rudder, and if there are crew injuries. Take the wind, waves, water depths, and all other pertinent circumstances into account before starting the tow. There should be an agreement on how the tow line will be passed, how it's fastened, who will man the helm, speed of the tow, the destination, and route to be used—before operations begin. Radios should be kept on a working channel and monitored throughout the tow.

Most skippers would choose to use their anchor line for towing operations. The elasticity of three-strand nylon anchor line is beneficial for this purpose because it reduces shock loads on both vessels. However, if the line breaks under load, it can snap back with incredible force and cause damage or injury. Two-strand nylon is strong with less elasticity, and is actually a better choice for towing.

Fig. 3.1
This polypropylene line is coiled fairly to pay out well.

Fig. 3.2
This Coast Guardsman has heaved a feeder line toward the stricken sailboat. The line will settle on the bow area.

Which boat supplies the tow line can be an important choice; this will be discussed later because of its legal implications. Passing the line to the other vessel can be done in a number of ways. In calm conditions, it can be as simple as approaching the disabled vessel from astern, close enough to pass the line over the side near the bow. When wind, wave action, or shoal waters prevent a close approach, heaving a smaller, waterproof, polypropylene line to the other vessel also works well (Fig. 3.1, above left). This is connected to the primary tow line, which is then pulled aboard with the smaller line.

The line is coiled carefully before the heave, ensuring that it pays out smoothly. A small weighted object is aimed over the vessel so that the line descends onto the decks (Fig. 3.2, page 26). Once this is captured, the tow line is attached and hauled to the other vessel.

When conditions make it more difficult, don't risk getting too close (Fig. 3.3, right). Maneuver to windward of the stranded boat, tie the tow line or a smaller line to a fender, and either heave it across to them on the foredeck or toss it in the water and let it drift toward the other vessel. The other crew can retrieve the line with a boathook. This may also work better with a smaller line first, depending on wind and wave conditions.

Whatever method is used to pass the line, the next step is to secure it correctly to both vessels. Figure 3.4 (page 28) depicts two vessels in position to begin towing operations.

Fig. 3.3
Care must be taken when approaching a stricken vessel in a seaway. Heave a feeder line from a safe distance, or allow the line to drift to the other vessel on a makeshift float.

Boat 1 has set up a bridle to which the towing line is attached. Depicted here at position A, it is attached to the two aft cleats. The bridle can also extend forward to stout cleats or encircle the entire boat to distribute the load. Its attachments should be easy to cast off; cleat hitches, made with several turns around the cleats, are

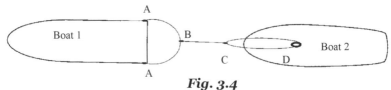

Fig. 3.4
Two vessels set with bridles and tow line.

recommended. A strong ring or snatch block for attachment of the tow line works nicely at the center point of the bridle.

From the towing vessel, the line may be attached to the bridle (position B) by a double bowline, anchor hitch, or any other reliable method. I prefer the double bowline because it provides two loops around the bridle, rather than just one loop of line.

The tow line should be long enough to allow adequate distance between vessels. This is determined primarily by sea conditions, the course to be taken, and size of the vessels (see following). It is usually best to allow for some sag (catenary) in the tow line, which dampens shock loading effects. Tows requiring greater protection against shock loading—resulting when the line suddenly becomes taut—can incorporate the stricken vessel's *anchor* in the tow. The weight of the anchor keeps some slack in the line, dissipating the loads encountered to protect both vessels.

The best method is to attach a stout shackle to a line from the towing vessel. That shackle is then connected to the stricken vessel's anchor, and its anchor line is secured on the stricken vessel. The anchor, thus suspended between the vessels, helps to maintain greater catenary in the line. This technique would usually be used in deep water where wave action threatened to hinder the tow efforts.

The crew of the stranded vessel (boat 2) also arranges a bridle to spread the load and strain of the tow. The bridle can simply be looped around the mast as in Figure 3.2 (which was the method we used aboard the Etchells), can utilize the forward cleats, or can incorporate the mast and forward cleats to further spread the load.

Whatever method is used to fashion the bridle, it is secured with a loop extending forward of the vessel for attachment to the tow line. Again, connecting to a snatch block or ring prevents chafe that occurs when the tow line is connected directly to the bridle.

If the tow line belongs to the towing vessel, it is most convenient to arrange for easy release of the bridle from that line when the tow is finished. Any hardware involved should belong to the towing vessel, so that releasing one end of the bridle from Boat 2 allows it to slip out of its connection with the towing line.

The line should be adjusted according to the situation—perhaps a boat length for a short-term tow out of a busy seaway or up a winding river, much longer in a tossing seaway. The objects are to prevent the tow from riding forward toward the tow vessel and to prevent excessive shock loading from too much slack in the tow line.

Fig. 3.5
This open ocean tow began with inadequate tow line deployed. The vessel aft is too close, and the boats are not synchronized with each other and the waves.

While towing in a seaway where waves and swells predominate, the key is to keep the boats in phase; when the tow boat is cresting a wave, so should the other vessel. Avoid the situation in which the front boat accelerates down a wave face while the rear boat is ascending a wave aft, creating great resistance that could suddenly tighten the line and jolt both boats. This will require a much longer tow line, which should be monitored for any changes in wave length and adjusted as necessary. The tow line should be shortened when the boats reach the calmer waters of a harbor to gain more control over the tow.

Fig. 3.6
The tow line has been eased out, creating more distance between vessels. This decreases shock loading on the tow line, keeps the boats synchronized, and reduces the possibility of the tow riding up toward the vessel giving assistance.

Courtesy of U.S. Coast Guard

When the lines are suitably secured on both vessels, radio communications established, and all other agreements are concluded, both crews should be alerted before the tow can proceed. The rule here is to begin very slowly, without a rapid tightening and jerk of the tow line. The towing vessel begins moving very slowly, gradually increasing speed not to exceed the hull speed of the tow. Speed is also governed by sea conditions, the course to be taken, damage to the tow, and so forth.

Crews on both vessels must be concerned with chafing of the towing bridles wherever they contact cleats, the mast, or any projection of the deck. These points should be monitored throughout extended tows, and the bridles released a few inches occasionally to prevent chafing. The chafe gear used for the sea anchor is ideal for this application. An old favorite for chafe prevention is old fire hose from local fire stations. Hoses are constantly replaced by newer models, and firemen have no use for the old hoses. Any adjustment of lines is only done when the towing vessel slows to allow slack in the line after communication between vessels.

Fig. 3.7
Both vessels are seen here in wave troughs, synchronized with each other and the waves. This can really be an eerie sight at night.

A plan must be in place to quickly release the tow rope from both vessels if something goes wrong. This could involve releasing the line from the bridles, or the bridles from the boat. Do not attempt to release the lines when under heavy pressure; the towing vessel should slow down to provide slack in the lines first. Have a sharp, serrated knife at the ready in case the lines have to be cut on short notice.

An anchor of the boat receiving assistance should be ready for immediate deployment if the tow line breaks in a narrow channel or near shallow water. When preparing to begin the tow, no crewpersons should be on the foredeck area of the salvage boat, or the cockpit of the towing vessel, except to manage the operation.

Keep weight aft on the boat in tow. This can control her tendency to yaw while under way, and keeps the crew clear in case of tow line failure. Someone should also maintain her constant heading behind the tow boat by actively steering. Control the speed and monitor the state of the tow line; this is no time to be in a hurry.

If the tow must traverse a narrow or winding waterway, or if traffic is heavy, decreasing the distance between vessels allows for greater control. This can be done by shortening the tow line to a minimum distance that is safe for the conditions. Sometimes, it makes more sense to tow the vessel "on your hip" instead of from astern. "Towing alongside," a common arrangement used by tugboats, makes the two vessels behave more like one to maximize control.

The lines used to connect the vessels are the same as those used for docking: a stern and bow line with two springs, as in Figure 3.8.

These lines should be tight to keep the vessels wedded together, with multiple fenders used between the boats. When

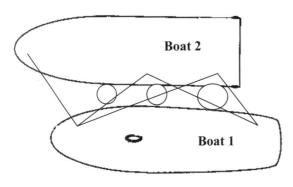

Fig. 3.8
The tow is lashed to the starboard side in a forward position. Towing alongside increases maneuverability and control.

two sailboats are involved, the aft position of the tow must prevent their spreaders from clashing. The stricken vessel (Boat 2 in 3.8) is usually lashed to starboard of the tow vessel, and maintains a forward position. This allows the tow boat propeller's normal action of driving the bow to port to compensate for the tow's drag on the starboard side.

Piloting the towing vessel in this configuration takes some getting used to; while the two vessels act like one, that one is slow and sluggish to control. The helm will have to counteract the bow's tendency to turn toward the other vessel, acceleration and maneuverability will be diminished, and it'll take longer to stop. Proceed at a safe, slow speed, and anticipate upcoming turns or other maneuvers.

A plan should be developed and communicated between the two vessels before the tow approaches a final destination, and crews on both vessels should understand that arrangement. The tow line to a vessel astern should be shortened to a minimum safe distance when in calm waters, entering a harbor or river, and when the operation is near a conclusion. It's usually preferable to release a vessel towed alongside to an astern position near the end of operations so that there are fewer lines to release when the time comes. The tow line should be readied for quick release, fenders and dock lines at the ready. It is usually wise to have a "roving" fender: a crewperson who watches carefully and places the fender where it's needed the most during final approach. The towed boat's anchor should be available for rapid deployment if needed. The easiest method of delivering the rescued boat is to place it on the outside of a T-dock.

The ideal approach faces into the wind/current if towing astern. The boat can tie up either to port or to starboard, whichever is most advantageous. The boat towed alongside must dock on its starboard side unless it is transferred to a stern position before release. Speed is slowed to bare steerageway that enables the towed vessel to coast into position when on final approach.

If the plan is to place the boat into a slip, she should be readied for docking as before. The boats approach into the wind/current if possible. The object is to release the tow with appropriate speed to enable a turn into the well with enough momentum to get at least the bow lines attached. Excessive speed at the point of release is far worse than too little. If the engine can't be used, there will be no way to slow down while entering the slip and the speed at release from the tow is more critical. After release, the vessel can be warped into position once it proceeds at least partially into the slip. It is then secured with the normal dock lines. This must be done with enough room for the towing boat to turn away and miss the outer pilings.

If the towed vessel's engine is operational, it should be started prior to the final approach, and used to bring the vessel into the slip.

If the operation runs past sunset or during restricted visibility (fog, heavy rain), both boats must display appropriate lights according to the *Inland Navigational Rules of 1980*.

Since sailboats aren't rigged with towing lights, the best option is for the towing boat to illuminate her running, stern, and steaming lights (assuming the engine is being used). The other vessel should display her running and stern lights. If her batteries are dead, flashlights can at least illuminate the decks, as can a spotlight from the towing vessel.

Dinghies are usually towed astern while cruising between islands or on short hops across a bay. There are other considerations for a dinghy at sea. Dinghies towed astern often tend to ride waves forward to overtake the boat, fishtail, and at times jerk about crazily. Adjust the tow line to place the two vessels in synchrony with waves; both should be ascending and descending waves at the same time. The dinghy should be at least one boat length aft—more with greater wave activity. Warps can be tied to the dinghy transom to prevent it from planing forward toward the boat if necessary. Excessive fishtailing is curtailed by lashing the dinghy with two lanyards, one to each quarter of the tow.

Dinghies are not towed at sea; instead, they're loaded and secured on board in the most practicable manner available. It's usually best to deflate and fold the dinghy as compactly as possible and stow it in a locker, if one is available. If not, it can be lashed onto the deck. Inflated dinghies can also be lashed, usually on the foredeck, but waves breaking over the decks can threaten even secure lashings.

Some monohulls and many multihulls have stern davits, and many skippers are sorely tempted to hoist the dinghy there for offshore transport. I don't allow that on my deliveries, and I recommend against it to anyone who will listen. After seeing waves break over the stern of a boat, and feeling the pressure with which the water hits, I know those waves could tear a dinghy away from the transom along with whatever it's attached to. Stow the dinghy before departure; don't even think about bringing it aboard from the davits during heavy weather.

Make sure the dinghy's engine is fastened tightly to the transom, with the fuel line closed. The gas tank should be lashed to the floor, and oars stowed. If the dinghy has a rudder, fix or lash it amidships.

Most skippers don't put a lot of thought into securing the dinghy while at anchor; the day's sail is over and it's time to enjoy the evening. That attitude changes once they've been awakened by the squeaking sound of rubber against hull--it's a lot like fingernails on slate. I like to haul the dinghy bow up onto the swim platform and tie it there to prevent the two hulls from rubbing together. I've also tinkered with extending the boom outboard, held fore and aft with the preventer lines. Use the main halyard through a block on the boom end cap, and hoist the dinghy bow up slightly at the bow. This is a little more work than just hauling the dink up astern, so I only use this method when I want to be more seamanlike.

The Legalities of Tow vs. Salvage

The simple act of accepting assistance from another vessel has the potential to result in a claim for inordinate compensation on behalf of the entity providing that aid. It is extremely important to distinguish the difference between marine **towage** versus marine **salvage** in these situations, to prevent being dunned for a salvage award when the intent was to accept a tow.

The admiralty and maritime law of the United States has recognized that **salvage** is the compensation to persons lending voluntary assistance to a ship at sea or her cargo that results in that ship and/or cargo being saved in whole or in part from peril. The purpose of this policy is to promote not only humanitarian rescue of life and property, but maritime commerce as well.

The salvor must establish three elements to have a valid salvage claim and be entitled to a liberal salvage award:

(1) marine peril

(2) services voluntarily rendered

(3) success, in whole or in part, with contribution to such success by the service rendered by the salvor.

This peril does not have to be "imminent or absolute"; it suffices that there is danger, either presently or to be reasonably expected. The degree of peril--whether slight, moderate, or severe—affects only the amount of the award, but not the entitlement of the salvor to be rewarded.

Many questions arise as a result of a recreational vessel's being driven aground. Admiralty law recognizes grounded vessels as being in a state of peril. The grounded vessel is exposed to the vagaries of wind, weather, and waves, and without further assistance is considered to be subject to further damage and eventual breaking up or sinking. It is sufficient if, at the time assistance was rendered, the vessel was stranded so that she was subject to the potential danger of damage or destruction.

Groundings in shallow water pose additional questions with regard to the existence of a "marine peril." Factors considered to be important are:

—stresses placed on the hull

—damage that is resulting, or could result from, pounding of the hull as it's "bounced" upon the bottom.

—the passage of time that increases the probability of additional damage.

—inherent danger to the people involved in attempting to extricate the vessel.

Each case involving assistance to grounded vessels must be decided on its own merits.

In cases in which a grounded vessel is not in peril, was intentionally grounded, is undamaged on a soft bottom, is capable of freeing herself without assistance at the next tide, or in which the act of assistance is out of convenience rather than necessity, the service to her is actually towage rather than salvage. Simple towage is a service based on the employment of one vessel to expedite the voyage of another when nothing more is required than the acceleration of her progress as a mere means of saving time, or for considerations of convenience. This is differentiated from a salvage operation, which implies that there was some degree of danger or of extraordinary assistance to the vessel.

The hallmark of "towage" is the absence of peril. An example would be a situation in which a sailboat, minus the availability of auxiliary power, while proceeding under sail in light airs without difficulty, requests towing assistance from a power vessel to expedite the vessel's return to her mooring in order to allow the passengers to meet an appointment.

Some courts have designated a service to be either "salvage towage" or "extraordinary towage" to mark the distinction between the physical acts of towage and salvage. Towage may actually be salvage when it is rendered to property that truly is in danger or where danger is to be reasonably expected. The context of each situation must be of prime consideration, taking into account factors such as the locale, distance, degree of danger, and degree of physical risk to the salvors. Context is very important in determining both the nature of services and the amount of compensation.

There is no question that salvage services cannot be thrust upon an unwilling vessel master or owner who refuses them. The question can arise, however, as to whether the vessel master or owner took affirmative steps to decline salvage services or implicitly agreed to them.

When no one is aboard a vessel exposed to a marine peril, the salvor is not considered to be a trespasser, and may proceed to assist the vessel and make a claim for a salvage award. Technically, an abandoned vessel, or derelict (Fig. 3.9), is one that was left by its crew without intention to return and without hope of recovery.

Fig. 3.9
The vessel in this photograph fits the definition of derelict, having been left by a crew with no intention of returning.

Even when a derelict is salvaged, the vessel remains the property of her owner until positive intent to discontinue ownership is established. The salvor who has earned the right to a salvage award has what is termed a high-priority possessory, preferred maritime lien on the vessel, which is not ownership. The salvor must care for the vessel and make reasonable efforts to identify and locate the owner. The salvor may file an action to foreclose his salvage, and obtain clear title to the vessel.

Whenever a situation involving possible or inherent salvage arises, a vessel master and party offering service should enter into an agreement with a signed contract. The vessel's master has the authority to sign a salvage agreement without special authority from the owner, who is bound by the actions of the master. However, if a master signs a salvage agreement after the vessel has been assisted, that agreement requires authorization by the vessel owner.

Such a document should clearly establish whether the vessel accepted salvage services, and should answer questions of security, arbitration, interest, and attorney's fees, or other terms. The contract may or may not specify the compensation to be paid in the event of success. The services in question cannot be termed salvage if the contract is signed before action is taken, because the services are then not rendered voluntarily but as the result of a preexisting legal duty.

In so-called "no cure, no pay" salvage contracts, in which the salvor will be paid only in the event of success, the salvor retains his status as a "pure" salvor and retains also his salvage lien. Most salvage contracts presented to recreational vessel operators and owners are "no cure, no pay." Salvage contracts are held to standards similar to those under common law, and may be voided in cases such as fraud, collusion, mutual mistake, misrepresentation, suppression of material facts, or compulsion.

Salvage awards are not based on any fixed percentage of the vessel's value, but rather on the peculiar circumstances of each case. The determination of salvage awards has traditionally followed the guidance of the Supreme Court Computation. This doctrine takes into account:

1. The degree of danger from which the vessel was rescued.
2. The post-casualty value of the property saved.
3. The risk incurred in saving the property from impending peril.
4. The promptitude, skill and energy displayed in rendering the service and salving the property.
5. The value of the property employed by the salvors and the danger to which it was exposed.
6. The costs in terms of labor and materials expended by the salvors in rendering the salvage service.

Establish an agreement before accepting the services of another entity whenever questions regarding towage vs. salvage, degrees of peril, and compensation come into question. This agreement is best sealed by signatures of both parties, preferably before witnesses.

Do not allow the party giving assistance to dictate all parameters of the assistance; have input into decisions regarding the method of towing, destination, route, and so forth.

Use your equipment for the operation—anchor line, anchor, etc.—as much as possible.

But ultimately, in the event that immediate action is required to prevent damage to property or risk to human life, only a fool would let legalities stand in the way of prudent action. There are plenty of lawyers to sort out the questions afterward.

4

The Iron Jenny

Written by Ed Mapes
with expertise from Blue Water Marine Engines

The simple fact is that engines are integral, virtually indispensable equipment on modern sailing vessels. They are relied upon for auxiliary propulsion and to charge batteries that power the myriad of electronics now utilized. Fuel is one of the commodities—along with water and electricity—that most skippers ration, but just charging the batteries often dictates anywhere from 1–4 hours of running time daily. Engines must be maintained and sometimes repaired to keep this vital resource operating properly.

Not everyone sailing the oceans is a diesel mechanic, but having a sound working knowledge of the power plant is essential. You need to know how diesels operate and understand their component parts and systems—including the maintenance requirements for those systems—and you should be able to perform basic service on your engine. Taking a diesel engine course is a sound investment for anyone sailing beyond VHF range of help, and all skippers should take the time to look over their diesel owner's manual to gain an insight into the workings of their own engine. Once you've identified the components, concentrate on the skills needed to maintain and service the engine.

Why do you need such intimate knowledge of the propulsion machinery? Because if it fails to start, loses power, races, or shuts down at sea, it's on you to figure it out. Sorting out and then curing a stalled engine in a dimly-lit compartment that reeks of diesel fumes on a rolling ocean (Fig. 4-1, below) is much different than working in a classroom scenario or even at your marina, but the job goes more smoothly when your familiarity with the workings of diesel engines grows.

Fig. 4.1
The author sorts out causes of this stalled engine in spite of a deep rolling motion at sea.

Routine engine monitoring and maintenance must be standard procedure at sea, so be sure you can perform the following tasks:

- Check oil, add oil, and change the filter.
- Closely monitor fuel usage, switch fuel tanks, and add fuel from jerry jugs.
- Check the closed water system coolant levels and add coolant.
- Examine all hoses for cracks, breaks, or loose hose clamps, and replace any damaged hoses. All hoses connected to the engine must be approved for this usage. They should be fire retardant and flexible to absorb engine vibration. Each point of connection must be secured with two 316 stainless steel hose clamps facing in opposite directions.
- Monitor and service the alternator brackets, water pump, and refrigeration compressor and their belts.
- Replace the primary and secondary fuel filters. Understand how to keep the filters and hoses full before restarting the engine.
- Check the transmission fluid and refill as necessary.
- Bleed the fuel system.
- Trace a blockage in the fuel line and clear the obstruction. Remove the fuel line from the primary fuel filter and blow backward toward the fuel tank, or insert a long wire to push the obstruction back into the tank.
- Service an obstructed primary fuel filter.
- Clean out the raw water strainer.
- Examine, tighten, and replace engine belts.
- Shut off the engine manually by stopping fuel flow at the tank valve, at the fuel pump, or by interrupting airflow.
- Start the engine manually if the solenoid fails: insert

an insulated screwdriver across the solenoid positive-to-positive terminals or jump the starter with a hot wire.
- Manually operate the gearshift in case of linkage failure.
- Understand the linkages of gearshift and throttle and how to make repairs.
- Manually operate the engine throttle in case of linkage failure.
- Use the raw water intake hose to pump water from the boat in case of emergency flooding.
- Inspect and replace the raw water impeller.
- Monitor and service the stuffing box/stern gland.

One of the great advantages of sailing vessels is: they can sail. All we need is the composure and ability to capitalize on that asset to make port without the engine.

I was once on a delivery from Tortola to Annapolis on which we lost the engine and all charge in the batteries after covering the first 360 miles of the route. With one crewmember injured after a fall during rough sea conditions and the forward berth sole under water from an unexplained leak, the decision was finally made to return to Tortola.

We reversed course around midnight, bringing the boat about between 15-foot waves and 30+ knots of breeze, along with our compass and handheld GPS (and 2 extra AA batteries) and my trusty sextant. This was the first opportunity I'd ever had to voyage under these conditions, and at first it seemed a tall order. What happened, though, was truly magical and stands out among the best of my oceangoing experiences.

Our crew of three had already hand-steered for three full days without the autopilot. The following morning the conditions had eased, with winds now a very manageable 18 knots from about

70 degrees off the starboard bow. The solution to the steering problem was at hand! Whenever winds are from before the beam with moderate seas, the boat can be balanced with sail trim to steer itself—and she did just that for almost the entire distance back to port.

During the next three days, we encountered squalls, calms, cold coffee, and tremendous camaraderie. We approached landfall during the sixth night out and hove-to offshore until daylight. At dawn the next morning we brought her into port under sail, did a U-turn, and settled gently against the dock to conclude one of the most memorable passages I've ever been on.

Losing the engine doesn't have to be the end of the world (though the world can admittedly be a lot easier when the engine runs). In the event this should ever happen to you, be sure to provision the following:

- Paper charts of the entire route, including ports of departure and arrival and possible intervening havens of refuge.
- Plenty of spare batteries for the handheld GPS, camera, and so forth.
- Sextant and navigational publications are excellent navigational backups for those so inclined.

Servicing an engine—or anything else—requires your knowledge and ingenuity, spare parts, and tools. In the incident above, it turned out that all engine spares had been taken from the boat by unknown persons in the week preceding this delivery, some time after my inspection had confirmed they were on board. Here's a suggested list:

Engine spare parts for extended cruising and passagemaking

(Courtesy of *Ready to Sail.*)

- Belts
- Hoses
- Hose clamps
- Engine oil, at least two gallons
- Transmission fluid
- Water pump
- Water pump impellers
- Starter solenoid
- Starter motor, except if a crank mechanism is available
- Alternator
- Fuel pump
- Fuel filters, primary and secondary
- Antifreeze, if in colder climates
- Injector lines
- DRIP-FREE™ or other stuffing box repair material
- Penetrating oil
- Gasket paper
- Engine gasket kit
- Gearbox oil seal
- Gasket compound
- Gas engine spare parts: points, condenser, rotor, impeller and pin, diaphragm, ignition spray, spark plugs, coil

This parts inventory must be supplemented with a tool kit enabling you to make whatever repairs are necessary. Picture yourself in your marina, where boat motion is negligible and the lighting is perfect. The engine has stalled, and you're attempting to sort out the problem. Even in that controlled scenario, everything you do is predicated on having the right tool for the job. Jury rigging and making repairs can be challenging enough, but they become even more difficult when you also have to jury rig tools.

The ravages of corrosion take a toll on tools, just like everything else metallic. Use spray penetrating oil occasionally to lubricate and protect all tools with movable parts or they'll soon be frozen in position and unusable. Needle-nose pliers frozen in the closed position aren't much good.

The causes of engine failure we see most often are clogged fuel lines, overheating, low oil levels, and air in fuel conduits. If you're able to service these problems, you'll be well on the way to competency, but don't stop your education with just these essentials. Knowledge is power, especially when it can get you out of trouble.

Clogged Fuel Lines

It pays to be serious about preventing fuel problems. Contaminated fuel will damage the tank, lines, and engine components, and ultimately shut the engine down. Replacing a fuel filter and/or reaming out a clogged fuel line is a messy job that interrupts passages and can leave the smell of diesel fuel on your hands, clothing, and throughout the boat.

Approximately 90 percent of fuel system obstructions are due to either organic matter or the by-products of algal contamination. This sludgy, slimy, acidic material is what most people refer to as "diesel fuel algae."

Bits of corroded metal are the other chief offender. Metal fragments can originate in your own fuel tank(s), the tank(s) of the station at which you pumped the fuel, the delivery truck, or a storage container at the refinery. Except for a corroded tank in your boat, this contamination can be prevented by filtering all fuel as it's pumped into your boat. After loading contaminated diesel fuel from marinas on the Chesapeake Bay and in the Caribbean, I use a trusty Baja filter (Fig. 4.2, below) with every fill-up.

Fig. 4.2
This Baja fuel filter has three very fine filtration elements.

Squirt a few gallons of fuel through the Baja, and then examine the uppermost filter. If debris has been trapped, either stop filling or run the fuel in slowly so that the filters can remove the contamination, or find another fuel source.

Microorganisms live in the interface between diesel fuel and water. A water layer is inevitably present in fuel containers, coming from a variety of sources:

- Water that settles out at the refinery when warm, newly-produced fuel cools down.
- Condensation within storage units and tanks as air temperatures fluctuate.
- Water that enters tanks via vents, sampling ports, or leaking seals.

Wherever the water comes from, it allows a microbial population to thrive in the fuel tank. The organisms inhabit the water layer below while feeding on adjacent fuel, and liquid hydrocarbon fuels are an excellent source of nutrients for these species. A secondary set of inhabitants can also attach to the tank walls, serving as a reservoir of organisms to contaminate freshly added fuel.

The ensuing explosion of microbes deteriorates the fuel quality and forms hydrogen sulfate, an acid that corrodes fuel tanks and piping. Other by-products lead to haziness, odor, and the sludge formation that obstruct lines and filters. Our goals to control tank microbes are to:

- Pre-filter all fuel loaded.
- Use fuel additives to counteract microbial growth.
- Keep the tank either full or empty whenever the boat is inactive for prolonged periods.
- Polish the tank before using the vessel after a period of inactivity.

We've discussed filtering all fuel taken on board. After the tanks are filled, instill a fuel additive to prevent microbial proliferation. Fuel additives are available at marine chandleries and engine suppliers. Keeping the tank full also reduces the chance of fuel/water interface development, especially when the boat will be inactive for a period of time.

The engine will sputter intermittently when debris begins to clog a fuel line or the primary and secondary filters. As the obstruction becomes more complete, longer interruptions of normal combustion, with resulting power loss, will occur; this is a warning sign that the fuel flow is impaired: it's time to shut the engine off. Sometimes a sudden obstruction (most likely from tank to primary filter) will shut the engine down with no warning. With the engine off in either case, the first step is to secure the boat by anchoring, drifting to a safe area, maneuvering to the starboard side of a channel, or using whatever means is most practical. Once the vessel is safe, start your investigations to determine the cause.

Obstructed Fuel Lines

These are almost always found at the fuel tank outflow fitting or in the fuel hose between the tank and primary filter. Shut off the fuel tank valve and remove the hose (usually a black, flexible fuel hose) from its connection to the primary fuel filter. Clean the hose with a paper towel, open the engine valve, and then blow on the hose very gently to find out if an obstruction exists; resistance is a positive test. If found, blow forcefully to expel the material. Diesel fuel doesn't taste good, and leaves your lips a bit tingly, but the tactic is effective and quick.

Another method is to ream out the hose with a long, thin object. Some fuel tanks have a filter screen at the outflow fitting, though, and this method risks tearing that filter and the fuel line itself.

Once the obstruction is cleared, replace the line and give the engine a try.

Obstructed Primary Fuel Filter

Shut off the fuel tank valve to prevent air from entering the hose. Place absorbent material beneath the filter. Cut away the upper section of a plastic water jug, leaving the handle in place; this is where you'll pour the contaminated fuel. Unscrew the filter from the threaded shaft that holds it from above, and gently bring the filter down and then away without spilling fuel.

If the filter can't be removed by hand, use a filter wrench. Other means useful in this type of operation are to secure a hose clamp (possibly two joined together) around the filter and then pull against it with vice grips. A knot in a small-diameter length of line that can be used to unscrew a seized fuel filter can be configured in the following manner:

1. Tie the ends of a short length of line together. The line must fit around the filter, yet be strong enough to withstand considerable tension.
2. Lay the line around the filter and lead one end through the loop.
3. Lead the first end around the filter once or twice more.
4. Pass this through the first loop one more time.
5. Pull against the first loop to tighten the strop around the filter.
6. Use a long-handled screwdriver or manual bilge pump handle to pull back against the first loop with enough force to turn the filter.

Experiencing this difficulty in removing the filter teaches one not to over-tighten the new filter when putting it in place.

Once the filter is removed, check the observation bulb at the bottom for the presence of water. Empty the contaminated fuel from the filter into the plastic container. This fuel can be transferred to plastic jugs with lids, and stored for disposal later. Open a new filter, make sure it has the thick gasket in place, and dab some clean fuel on the gasket. Now fill the new filter all the way to the top with fresh diesel fuel and hand-tighten it into position. If no air was introduced into the fuel lines, bleeding the engine shouldn't be necessary.

A fouled filter can be cleaned for future use if necessary. Fuel filters have openings on both ends, so begin by covering the far end with a plastic bag, held firmly to the bottom. Next, half-fill the filter with clean fuel, and cover the top with another bag. Now, shaking vigorously will remove most of the debris from the filter element; dump out the contaminated fuel and store the filter to be used as a temporary fix later if needed. It should be replaced at the first opportunity.

After relieving a clogged line or replacing a dirty filter, open the tank valve and start the engine.

Fouled Secondary Fuel Filter

Secondary filters are located downstream from the primary filter and engine lift pump. They are often fitted very near a primary fuel pump that directs fuel toward the injector lines. This is the last line of defense in preventing material from entering the combustion chambers. Fouling of the primary filters is far more common than fouling of the secondary filters, since most contaminants are removed before making it this far. Minute bits of debris do elude the primaries, though, and over time will accumulate to obstruct fuel flow.

Sludge can also accumulate quickly if the fuel tank is "polished," which will dislocate from tank sides and bottom large quantities of algal material that can be aspirated into fuel lines. All of the material may not be removed by the polishing filtration, and some of it gets returned to the tank. The primary fuel filters are often changed after this operation, but not the secondaries.

Whenever the primary filter has been changed several times, there has obviously been considerable contamination of the fuel; the secondary should be inspected as a wise precaution. The gradual buildup of debris in the secondary filter can finally take its toll, causing engine shut-down.

Fuel interruption shuts down an engine whether it occurs at the primary or secondary filters. We naturally check the primary filter and fuel line first when the engine sputters or quits. If no obstructions are found there, though, turn your suspicions to the secondary filter before starting to pull out hair.

Overheating

The engine has purred for over 60 hours. We're motor-sailing toward landfall, still 300 miles off. With little to do except watch the autopilot steer the course, we hear the night's peace and stillness unexpectedly interrupted by the shrill sound of an engine alarm. You know this is usually the onset of a significant problem. Now what?

Diesel engines will run for thousands of hours if the fuel is clean, oil is maintained, air is plentiful, and cooling systems are functioning properly. In this instance, your first suspicion should be an overheated engine because it would simply shut off without fuel or air. Alarms may also sound with low oil pressure, but your actions should be the same no matter the cause: shut the engine off as soon as the boat is safely positioned to do so. Overheating chars components, melts head gaskets, destroys plastic mufflers, and eventually seizes the pistons.

A quick glance at the engine room bilge and dipstick rules out oil leakage or low oil levels. In this case, since the engine had been running so well, a stuck thermostat is unlikely.

If your inspection finds normal oil levels, skip the thermostat for now and begin the diagnostics of a failed cooling system. Inspect the coolant levels of the closed cooling system to make sure they're nominal; if so, the problem relates to the raw water side of the system, and a methodical approach must be used to find the problem.

Water enters the intake seacock and travels to the raw water filter (Fig. 4.3, page 54, left), which strains out particulate matter such as seaweed and leaves.

A hose exits the filter and proceeds to the raw water pump Fig. 4.4, page 54, right).

Fig. 4.3
Matter (such as seaweed, leaves, and basketballs) is removed from seawater in the raw water filter.

Fig. 4.4
This pump receives seawater from the filter and propels it out toward the heat exchanger.

Water may then be taken to other components, such as the refrigeration condenser, but ultimately it ends up in the heat exchanger. There it removes heat from the engine coolant much as a car radiator does, and then the heated water is usually delivered to the muffler. The water filling the muffler reservoir serves to quiet the exhaust noise it then empties into an exhaust hose that exits the boat through a port at the transom.

Begin the examination at the raw water intake seacock. Close the seacock valve handle, remove the water hose, and open the handle enough to check water flow. Plastic bags and other debris can be sucked into the intake and obstruct water flow. I once had to clear the seacock of Sargasso weed we picked up in the Gulf Stream. You can clear it by pushing the obstruction away with a long screwdriver or coat hanger. If no obstruction is found, replace the water hose and re-open the valve.

Next, inspect for obstructions—dirt, leaves, etc.—within the intake filter, and clean it if necessary. If you remove the lid, look for the O-ring. There must be an airtight seal that maintains a vacuum within the filter, and a defective or missing O-ring can be at fault. Replace a missing or damaged O-ring; coat it with Vaseline while replacing the lid, and take great care to screw it on properly. If the filter checks out, trace the hose exiting the water filter (Fig. 4.5, opposite page) to the water pump, which features a flat face plate secured with four to six screws. Before taking the plate off, close the intake seacock, remove the intake water hose from the pump and elevate it to prevent water from leaking out.

Impellers consist of a hub that fits over and meshes with the pump shaft. Pliable fins that normally terminate in hemispherical knobs emanate from the hub. These fins force raw water from intake to the pump's outlet ports.

Fins should show no distortion, rigidity, or cracks. They are subject to deterioration with age and normal usage, but fall apart rapidly if water is blocked upstream. Intact fins can also be shredded by broken fins trapped inside the pump. (Fins can stick to the pump walls after winter storage, and break apart when the engine is started in the springtime.)

I make a point of changing the impeller once every year and again if the engine has run a lot during a passage. I save used but intact impellers for possible use at another time.

If rotor fins are missing, remove the impeller by inserting a screwdriver behind the hub to gently pry it out (it's also possible to pull the impeller directly out using small pliers, but that can damage fins on a good impeller). You'll have to replace the impeller, but the first concern must be with the missing pieces. They can cause an obstruction within the pump (most likely) or downstream in the refrigeration condenser or heat exchanger.

Remove the hose that is **lowest** on the water pump; that should be the exit hose. You'll see several narrow slots (as in Fig. 4-5, page 56) in the pump housing. Shards of impeller fin can be lodged in

Fig. 4.5
The water pump on this 50-horse Yanmar has had the face plate and impeller removed to reveal its shaft and interior. Note the slots in which debris can become lodged when impeller fins are dislocated.

the slots; these are customarily removed by pushing through the slots from the outside of the pump with a short piece of wire, but there is an easier, faster, and more thorough way:

Reconnect the hose at the pump, and then remove the hose from its attachment at the water filter. Take a big breath and blow forcefully into the hose; this forces the obstructions out. Use needle-nosed pliers to remove all the visible shards, and then repeat the procedure and inspect the slots again to ensure a clear pathway. Collect all impeller shards and try to determine if they've all been accounted for.

Grease the fin tips of a new impeller with lubricant. Small tubes often come with new impellers, but you can use winch grease instead. Now place the replacement impeller over the geared pump shaft and twist **clockwise**— so that the fins bend counterclockwise—as you slide the impeller onto the pump shaft.

Next, reconnect the hose and make sure the system is primed with water from seacock to water pump. If water has leaked out, refill either at the filter or in the hose connecting to the pump. Water will not pump unless there is a vacuum throughout the system.

When you're sure it's primed with seawater, start the engine and watch the external port for water pumping through to the exhaust.

This should only take a few seconds, just until the muffler fills and creates enough back pressure to force water out the exhaust.

Had water still not pumped, our next procedures would be to observe that the impeller (most are belt-driven) turned inside the pump with the engine running and to examine for additional pieces of impeller fin in the water lines from pump to heat exchanger. Finally, we would examine the exchanger itself. In my experience, most impeller pieces have been located as described earlier, and water flows when those pieces are removed and a new impeller is installed.

If an engine with several years' usage overheats after performing all of the above, consider the water pump itself as the culprit. I encountered this situation with *Voyager*, which overheated on two passages even though I meticulously changed impellers, changed all water hoses in the system, and replaced the O-rings of the water filter and water pump. The pump looked fine, but replacing it solved the problem.

If the engine shuts down repeatedly after a few minutes of running time, odds are the problem is the thermostat. The thermostat is located within a bell housing near the front of the engine, and will have water hoses entering and exiting.

Once the engine has heated to optimum temperatures, the thermostat should open to allow raw water from the impeller to proceed into the heat exchanger to begin controlling the engine temperature. If the valve fails to open, the engine is not cooled, and overheats.

If the thermostat fails to operate, heated water entering from the engine will not open the valve, which should allow the water to enter the heat exchanger for cooling. You can bypass the thermostat and try running the engine without it to identify that as a problem, but that would be considered only after exploring more likely causes of the trouble.

While at sea, unless you've stowed a spare thermostat, the best fix is to bypass it altogether; connect the water hose from the impeller to that of the heat exchanger. This should cool the engine to normal levels. The thermostat should be replaced at the first opportunity.

Low Oil Levels

Proper lubrication is very important because diesel components fit together with such close tolerances. Heat builds up quickly without normal engine oil circulation. Before heading out for any extended sailing venture, and at intervals recommended by the engine manufacturer, change the oil and oil filters. Stock enough oil for a complete oil change, with a few quarts extra in case the oil plug falls out. Monitoring oil levels at sea is normally done before starting, at least twice daily, or every four hours when the engine runs continuously. Note that the engine must be shut off while checking the oil. Check the dipstick level and look under the engine for leaked oil, and refill when the levels drop. Be careful not to overfill, however; the increased pressure will shut the engine down or lead to engine race.

Bleeding the Fuel System

Combustion in diesels is achieved by the compression of a very precise air-fuel mixture. This mixture is aerosolized, and is then admitted into the cylinder at precisely the right moment by the intake valve. The mixture's components are then heated to the point of ignition by compression created by the piston's down stroke. That burns the fuel, creating the engine's power and torque. Extraneous air finding its way into fuel lines alters the air-fuel mixture and diminishes compression, which can decrease engine efficiency or cause outright shut-down. There is no alarm and usually no warning; the engine just stops. Unwarranted air entry is often the result of a leaking fuel hose, but can also be traced to the primary fuel filter, faulty fuel tank selection valve, or a valve left open or following work done on the fuel lines.

Check for the problem and remove air from the lines in the same operation. Trace the fuel line that exits the primary filter; it leads first to the lift pump. Note the toggle handle on this pump. Continue following the fuel line to the fuel pump, and locate a prominent bolt there, usually on top and in front. Loosen this bolt, and begin operating the lift pump handle to propel fuel through. Hold a paper towel near the opened bolt, observing for fuel and bubbles to emerge around the bolt. Continue pumping until only fuel streams from the bolt, and then tighten it with a wrench. Proceed to the injector housing and open another bolt there. The fuel can be propelled either with the lift pump or by a mate operating the engine starter. If no bubbles but only fuel appears, tighten the bolt; if you do see bubbles, continue pumping until only fuel streams out and then tighten the bolt.

Once done, start the engine and observe it running for a couple of minutes before resuming your course. Don't be alarmed if the engine fails to turn over immediately; be willing to use the starter for a few seconds.

The problems outlined above cause the vast majority of engine disorders, but by no means are they the only ones that arise. The more thoroughly we understand diesels' components and operation, the better equipped we'll be to maintain efficient operation. Be vigilant in engine maintenance before and during a passage, but don't forget the big advantage we sailors have: use the wind when all else fails!

SAFER OFFSHORE

5

Rigging Failures

Written by Ed Mapes
with contributions
from Mike Meer

Watching helplessly as a mast topples to the decks on an inland lake is one thing, but losing the rig a few hundred miles from land on a storm-ravaged ocean or while attempting to claw off a lee shore, are different matters. Left unattended, rig defects can insidiously escalate into life-threatening failures when put to the test by heavy weather (Fig. 5.1, following page). These are situations that our on-shore inspections and monitoring at sea aim to prevent, but sometimes even the best of efforts go unrewarded. That's when jury rigging talents come into play.

This chapter will begin with a discussion of how rigs fail, emphasizing the importance of constant monitoring to detect early problems with the standing and running rigging. This should enable us to make repairs before defects cost us a spar. Later the discussion turns to salvaging what's left after rig trouble, enabling you to make landfall.

Fig. 5.1
***The power of wind and waves took a heavy toll
on this merchant vessel, washing it ashore with heavy losses.***

Please see Chapter 3 in *Ready to Sail* for a thorough discussion of the systematic inspection of your rig that should be carried out before leaving port. This method has helped me, delivery skippers, and hundreds of sailors find and correct rig defects for over thirty years. More than a few of these problems would have been very costly if the craft had taken to sea unchecked.

On the facing page is the inspection checklist from Chapter 3, "Spars, Rigging, Sails, and Canvas", of *Ready to Sail, Volumes I and II*. It illustrates the vast number of components involved in this system, and sheds light on the multitude of failures that can spell ruin.

The mast(s) and standing rigging are such fundamentally important structures that no effort should be spared in their upkeep. A lackadaisical attitude toward maintenance is the single most important reason for rig failure. With normal usage, exposure to the elements, and the passage of time, components eventually wear

Ready to Sail

SPARS, RIGGING, SAILS AND CANVAS

- Sheets
- Downhauls
- Topping Lifts
- Shackles
- Feeder Line
- Cheek Blocks
- Turning Blocks
- Fairleads
- Main Traveler
- Mizzen Traveler
- Line Stoppers
- Winches
- Mainsail
- Jibs
- Mizzen
- Staysails
- Spinnakers
- Other Sails
- Telltales
- Battens
- Reef Lines
- Forestay Slide

- Roller Reefing Drum
- Head Swivel
- Shrouds
- Forestay
- Backstay
- Running Backstays
- Chain Plates
- Turnbuckles
- Shroud Eyes
- Stay Eyes
- Main Reefing
- Boom
- Boom Vang
- Outhaul
- Bails
- Mast Bolts
- Mast Boot
- Wiring
- Partners
- Mast Position
- Spreaders
- Steaming Light
- Deck Light

- Anchor Light
- Strobe Light
- Tricolor Light
- Baby Stay
- Spinnaker Track
- Mast Tangs
- Mast Head
- Mast Head Sheaves
- Antennae
- Wind Instruments
- Windex

Notes:

Courtesy of *Ready to Sail*

RIGGING FAILURES

63

Fig. 5.2
This stem ball failed with no warning.

Fig. 5.3
The upper rod terminal is cracked beneath the telltale rust.

out—sometimes presenting little evidence of impending breakdown (Fig. 5.2, left).

The first part of any rigging inspection involves an investigation into the boat's history. Age, location, and previous use all play roles in the assessment of the spars and rigging. Recently Navtec, a leader in stainless steel standing rigging systems, published a service manual outlining the maintenance intervals of various components. A general consensus among rigging professionals is that the useful life expectancy of 1 x 19 standing rigging is around 15 years. This number drops significantly as the salinity of the water and average temperature increases; a boat on the Chesapeake Bay gets to the 15-year mark, but the same rigging in the Caribbean lasts a mere 8 years. Previous use should also be assessed. A boat that has been to sea and sailed hard is more suspect than one that has seen only light duty. Be warned, however: a boat that has spent her whole life sitting at the dock is also cause for concern. The static loading of an idle rig can cause a host of problems, corrosion and point loading of clevis pins being the primary issues.

It makes sense that the greatest wear takes place where connecting components are damaged by friction. Toggles and their clevis pins, terminal fittings, mast attachment points (tangs), spreaders, and chainplates are key areas of concern. Our inspections must be careful and vigilant; while defects in this system are potentially catastrophic, they're yet often subtle.

A principle cause of rigging failure is the motion of stays and shrouds. When the top and bottom of the system are not toggled to allow freedom of fore-and-aft and lateral motion, undue stress (motion fatigue) is placed on the attachment points at both ends. This results in erosion, deterioration, and cracked metal components (Figs. 5.4 and 5.5, right). A second toggle that allows lateral motion—at the top and bottom of the rigging—is the key to eliminating motion fatigue. The toggle in Fig. 5.6 (next page) illustrates the ultimate result of motion fatigue.

Alignment of the standing rigging and fittings to the direction of the load is also very important. Misalignments caused by interference or bends cause increased local stress and dramatically reduce the working life of the rigging. Any kinked, bent, or improperly aligned wire, rod, or fitting should be replaced, because the damage is permanent.

All components of the system should fit together correctly to avoid uneven wear of bearing

Fig. 5.4
The single toggle shroud arrangement permits motion in only one direction. The clevis pin is also badly corroded.

Fig. 5.5
The perpendicular toggles alleviate motion fatigue in Voyager's *inner forestay.*

Fig. 5.6
End result of motion fatigue and corrosion. Here, a toggle that allowed motion only in the lateral plane finally succumbed, and the shroud was lost.

surfaces. Toggles must be correctly sized according to the chainplate or mast fitting, and clevis pins have to fit precisely to prevent untoward motion and bending. Clevis pins that don't match the holes in the chainplate toggles can deform the holes or cause bending or cracking of the pin. Another leading cause of rig failure involves the most common wire terminals, swage fittings (Fig. 5.8, opposite page).

Swage fittings are crimped over wire under tremendous force, forming indentations within the fitting as it conforms to the shape of the wire strands. These fittings are at risk from wear and corrosion, predominately at the lower ends, because of exposure to the salt and pollutants prevalent in seawater. Poorly executed swage fittings are also cause for concern. A banana-shaped swage causes load to be placed unevenly across the wire.

Fig. 5.7
Wear and old age have caused an elongation of the pinhole of this toggle, indicating imminent failure.

Courtesy of Southbound Cruising Services

The braided nature of wire encourages the infiltration of water and salt that cause corrosion, especially at the junction with swages. Plastic covers designed to protect sheets from chafing on shrouds prevent the wire from drying out, accelerating corrosion. Over time, the swages deteriorate, developing minute cracks that lead to failure when the wire suddenly pulls out of the swage.

Fig. 5.8
Corrosion is rampant on this upper swaged fork terminal, and deterioration is well along.

Fig. 5.9
A single broken strand in 1 x 19 wire. The shroud must be replaced.

One-by-nineteen stainless steel wire is best suited for shrouds and stays on cruising vessels. It is more flexible, lasts longer, is easier to install, and costs much less than rod rigging. In addition, stainless steel wire gives some early indication of fatigue and corrosion, usually displaying separated, bent, rusted, or unwinding braids of wire.

Corrosion, pitting, and cracks—especially when perpendicular to forces on the component—are mainly what we look for. Cracks can often be found by close visual inspection, but you can also use a magnifying glass or dye penetration tests. Professionals also employ X-ray, ultrasound, and eddy current testing.

The wire or rod and all fittings must be clean before visual inspection to uncover hidden defects. A reliable telltale sign is the presence of rust, which often indicates cracks underneath. Any areas showing discoloration or potential corrosion should be thoroughly cleaned and examined for further damage.

If you're at sea, you can evaluate the upper mast structures and connections by using binoculars, but when you're at the dock, you should take a trip aloft for a thorough look. Up high, shrouds are connected to spars in a number of ways, and we should be very familiar with the system used on our boats to effectively spot defects (Fig. 5.11, below).

Fig. 5.10
Bending causes local weakening and misalignment of the shroud.

A   B   C   D   E

A) Single clevis tang
B) Double clevis tang
C) T-ball backing plate tang
D) Round Stemball Backing Shell
E) Rod stemball tang

Fig. 5.11
An assortment of mast fittings.

Courtesy of Rig Rite

The great loads on these structures and the ravages of corrosion add to the likelihood of failure (Fig. 5.12, right top).

Many attachment points on masts are welded in place. This isn't as foolproof as you'd think, though, as even these can fail (Fig. 5.13, right bottom).

Masts can break in a number of locations, depending on the forces exerted versus those that keep the spar in position. Should the boat turn 360 degrees after being tumbled by a wave, the whole mast can be torn off, especially if it's deck-stepped. A break can occur in the upper section if a D2 shroud fails. The most common site of mast failure though, is just above the lower shrouds, where maximum bending takes place, as in Fig. 5.14 page 70, top left). Incorrect shroud and excessive backstay tensions are frequent contributors, along with component failure.

Fractures in the mast can be subtle (Figs. 5.15, 5.16 page 70 top right) or suddenly disastrous (Figs. 5.17, 5.18, 5.19, page 70 bottom), depending on the causative factors and immediate remedial action taken.

Fig. 5.12
A dangerous crack is seen at the top of this double clevis tang.

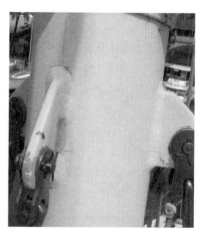

Fig. 5.13
Broken forward diamond stay attachment lug.

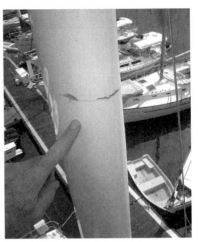

Fig. 5.14
With maximum mast bending, the area just above the lower shrouds is a prime site of mast failure.

Fig. 5.15
This incipient mast fracture was detected on predelivery inspection. The owner had been unaware of the defect, which necessitated extensive repairs before I took the boat to sea.

Fig. 5.16
Decking at the mast base is cracked because of unequal shroud tensions and insufficient mast base chocking.

Rigging Failures

Fig. 5.17
This spar fractured at the point of shroud attachment. Note the stem ball fitting that leads to the supportive rod.

Fig. 5.18
Catastrophic multisite mast failure.

The proper alignment of spreaders is also fundamental in rigging longevity. Spreaders are flattened on the horizontal plane, with great tensile strength forward but comparatively little vertically. Most spreaders angle aft and tilt upward a few degrees to bisect the angle of the shroud and maintain correct shroud positioning. The inboard connection is securely bolted into mast indentations, fastened to external brackets or fitted over through-bars that run perpendicular all the way through the spar. Outboard, tips secure the stay to the spreader and hold the spreader in the proper orientation. The connection can be as simple as a recess in the extrusion secured with monel seizing wire (using monel is critical because it does not cause dissimilar metal corrosion and it does not work-harden like steel) or a more complicated bolt-on tip. Either way, check for corrosion and ensure a tight fit.

Spreader damage can result from entangled sheets and sails, but accidents also occur during boat hauls, mast stepping, and collisions. Whether the tips, main extrusion, or the mast connection is compromised, any damage to spreaders can allow the spar to become out of column, with possible disastrous results. Every inspection must include these spreader areas to identify flaws and defects early.

Unfortunate events in the 1979 Fastnet Race illustrated mast failure where no one really expected it: at the mast base. The step (or sleeve) is a metal structure attached very securely to the keel or deck. Lateral motion of the mast butt is prevented by its precise fit into this sleeve. Experiences during the Fastnet, though, demonstrated that the sleeve does nothing to prevent a mast from moving *upward* off of the step. When standing rigs fail, the mast is free to pop off, right out of the boat (best case) or be driven into the boat and through the hull (obvious worse case). Installation of a large bolt, keeping the mast within the sleeve, is an easy fix for this eventuality. On *Voyager*, (Fig. 5.19) I drilled and tapped a hole for a 5/8" stainless steel bolt through both the sleeve and mast.

Fig. 5.19
This bolt through the mast sleeve into the mast prevents upward motion of the mast, even in the event of standing rig failure.

Mast integrity depends on more than shrouds and stays; running rigging is a factor in strength as well as performance. Although defects here wouldn't usually cause rig loss, they can be dangerous nonetheless. This system consists of all halyards and sheets, along with every component through which they lead, turn, or stop. Fig. 5.20 (right) illustrates a cracked mainsheet traveler toggle about to fail.

That calls into question how well sheaves, blocks, clutches, tracks, and winches are fastened to the boat, and their functionality. A Pad eye that pulls out of the deck or a seized masthead sheave will lead to trouble that is sometimes hard to get out of. A cracked snap shackle, for example, could free a spinnaker from its halyard or suddenly release a sheet from its sail.

Fig. 5.20
The boom will be uncontrolled when this fitting breaks.

Always inspect after others make repairs or after gear is added to your boat. Owners should try to do their own work if possible; that eliminates potential poor workmanship done by others and helps to improve overall boat knowledge. Figs. 5.21 (page 74 top) and 5.22 (page 74 bottom) demonstrate this principle.

Chafing of lines—from age, exposure to elements, or inappropriate wear in a particular location—is a constant threat to running rigging. Lines tend to chafe most at turning points--mast head and boom sheaves are prime examples--and where they come in contact with abrasive components. A great example of sheave chafe is the boom topping lift, which often remains in the same position over the mast head sheave for months at a time. This is often missed by boat owners: out of sight, out of mind.

The forward boom exit point of reefing lines deserves special attention. The upper edges of boom toggles are sometimes sharp, often chafing reefing lines that are pulled taut over them.

Fig. 5.21
The replacement boom toggle is clearly too small for this boat, as exemplified by the stack of washers used to make it fit. It is designed to accommodate a smaller working load, and is inappropriate and dangerous on this larger rig.

Fig. 5.22
This spinnaker block is improperly positioned. The halyard will chafe fast on starboard tack.

Chafing happens when the mainsail is lowered during reefing; the forward bull ring (cringle on smaller sails) and reef line pulls down against the boom, and the reef line crosses directly over the boom cringle edge. I have witnessed this chafe through the line during one reefing event. Prevent this by only lowering the main sail to within one foot of the boom initially, and then hauling in the reefing line. As the sail is reefed, you can carefully lower the halyard more, but don't allow the bull ring to approach the boom.

This is, incidentally, the main reason to carry an electrician's wire fish in the rigging repair kit. When reef lines break, replacements are tough to feed through the boom.

Chafe and aging of the running rig is a slow process that's detected by simple monitoring. Affected lines become weakened and more apt to give out under the strain of sailing. Chafed ends can be cut off, and the new terminal burned or whipped to keep the line serviceable. A line chafed more toward the center should be replaced. Aging lines become less pliable, and are coarse to the touch. This is especially noticeable while coiling them. Aged, dried line is also larger in diameter than new line, and may become more difficult to pull through blocks or clutches. Even though old line may not *appear* frayed or weakened, it is, and should be replaced. If you're like me, you have lots of lines on board salvaged after their replacement with newer ones.

Always have two jib halyards, or one active and one leader placed through the mast and over the masthead sheave for quick replacement if the primary breaks. The same holds true if you sail with spinnakers. Additional halyards are especially important for going aloft on vessels with furled jib and mainsails; those halyards are always in use.

Another troublesome problem occurs when sheets wrap incorrectly around winches, causing overrides that are difficult to sort out. Whenever tailing a sheet for a winch grinder, pull slightly upward on the sheet to prevent such wraps. Also, be sure that all slack is out of that sheet before the opposite sheet is loosened during a sailing maneuver.

On occasion a wrap can be cured by pulling upward sharply on the sheet as it circles the winch. If the override isn't too deep within the wraps, it may pop out. At times, the best option is to relieve all tension on the sheet, and unwind the wraps until the override is cured. This is most quickly done by tying a rolling hitch onto the sheet before the winch, and then leading that line to another winch to tension the sail and relieve tension in the primary sheet. Once pressure is taken onto the replacement sheet, the override is easy to unwind.

Headsail furlers can be the source of problems as well. Many a broken headstay has been caused by "halyard wrap," which occurs when the jib halyard twists around the uppermost foil, binding the furler and wrapping around the headstay. It is common practice among novice sailors to put the furling line onto a winch and start grinding in an attempt to free the jammed furler; the additional purchase can break the headstay at the masthead leaving the halyard and luff tape of the headsail to hold the rig up.

Halyard wraps can be avoided by ensuring that the jib halyard does not run parallel to the headstay for more than a few inches. The halyard should lead *aft toward the mast*, rather than following the path of the headstay. A restrainer attached to the mast aloft can deflect the halyard away from the headstay. The addition of a pennant, affixed to the sail, raises the top swivel, thus shortening the halyard length, to reduce the tendency to wrap. Another way to raise the swivel aloft is to add shackles or a small pennant to the bottom of the tack or on the head of the sail, as long as it raises the halyard swivel. Actually, placing the pennant at the bottom may improve visibility beneath the sail.

It is important that whenever undue tension exists on the furling line, you STOP furling and investigate the cause. Most commonly it is a wrapped halyard.

Whenever the vessel is sailed over great distances with a partially furled genoa, it is wise to let it out a couple inches every twelve hours to minimize chafe of the furling line. That prevents the line from breaking and suddenly leaving you with a full genoa, at the worst of times.

"You can't control the weather once you're out there, but you'd better control the weather you leave in." Safe skippers have been living by these wise words forever, and there are countless examples of trouble encountered by those who ignored them. Things go wrong at sea. It's a truism we have to accept. But it's only made worse when we sail on boats with unseen defects just waiting for the right situation to fail. It has to be standard practice to inspect the boat and its gear before taking to sea. While this is not a guarantee against trouble, every problem (Fig. 5.23, below) that we remedy beforehand is one less we'll have to cope with later, along with other mishaps stemming from that one.

Fig. 5.23
This double toggle chainplate holds only one shroud, but the bolt has almost unscrewed. The rig will topple if this shroud lets go.

Once we've gained the open ocean, it's up to us to manage the vessel in all regards, and the rigging is a vital component. Please refer to the Monitoring and Maintenance schedule in the Appendix. This is what I use offshore to ensure continued vigilance. The standing and running rigging are checked at least daily, every day we're at sea. What we can plainly see, we eyeball, what's farther aloft we go over with binoculars. Evidence of chafe to any line, damage to sails, or any part of the standing rigging is addressed immediately, before the condition has a chance to worsen--which can happen very fast when the wind blows hard.

It's not uncommon to be on the same tack for days on end at sea. I've been on entire offshore passages without a single tack or jibe, just trimming sails on the same board until landfall. This constant deployment and continual strain on the same components accelerates wear and chafe, and calls for careful monitoring.

Any chafed or cut lines should be replaced immediately. Look for whatever caused the damage as well, to prevent reoccurrence if possible. Jib sheets are likely candidates in the above scenario, chafing on shrouds, around blocks, on the forward components of lifelines, and so on. Don't forget the defective snap shackle discussed earlier.

If a line can't immediately be replaced, there are ways to manage temporarily. Tie on a substitute line using a rolling hitch, and transfer the load to the new line. If chafe is in the middle of a bearing line, a sheepshank knot will transfer the tension away from the weakened segment.

Most halyards terminate with a spliced-in eye that accommodates a suitable shackle. If a halyard eye should chafe or break, we have several alternative fixes. The bowline is probably used most often for tying the line on directly. I like to connect lines with two loops onto whatever appliances it holds—shackle, bridled line, etc.—when the loads are large. This can be done with the bowline, passing two loops instead of one (making it a French bowline), but my favorite attachment is an adaptation of the hangman's noose. It's tied by passing the halyard end through the shackle twice for the two loops, and then making successive loose turns around the standing part. After 4-5 turns, pass the end back upward through the loops just made. Maneuver the knot down toward the shackle, and pull it tight. This works well because it ends with the knot very close to the object and it tightens when tension is applied.

I've recently added an item to my rigging repair kit: the Clamp-Tite tool. This handy gadget enables one to form new eyes in line by very tightly securing stainless steel wire around it. The tool can be used for repairs in a number of applications, from rigging to plumbing.

The mainsheet traveler system involves a boom bail, the block and tackle portion, the traveler, and the car with its control lines and blocks. Should mainsheet control be lost for whatever reason, we must construct a replacement to control the boom. The most common failures are boom bails and weakened sheets. I have double-looped a stout line around the boom and formed a double loop with a French bowline, to which a shackle was used to attach the mainsheet double block. Never be afraid to tie lines completely around the boom; this provides the strongest attachment.

Another example of tying lines around the boom is in fabricating a preventer system. Tie both ends of a bridle around the boom, one several feet from the outboard end, the other just forward of the center. Install pad eyes on either side of the boom to anchor the bridles in place, one to port and one to starboard. Preventer lines attach to the bridles as in Fig. 5.24, page 80.

I prefer to configure a boom bridle, as opposed to using a single point of boom attachment, since it spreads the load along a greater area. I once saw a carbon fiber boom demolished because the preventer was attached to the outboard end. The boat was rolled by a large wave, snapping the boom in its middle section. The bridle protects the boom from bending under such heavy loads.

Voyager's preventer system, as in Fig. 5.24, consists of two preventer lines, each connected to its own bridle on the boom. Notice that stout shackles (rings can also be used but make adjustments of the lines more difficult) are attached in the center of each bridle. The actual preventer lines are attached to these shackles, and are then led to blocks along the deck positioned to pull the boom forward. From these blocks the lines are taken back to the cockpit for easy control.

Fig. 5.24
Voyager's preventer.

Once the course is set and main sail trimmed on *Voyager*, I secure the leeward preventer line to the large aft cleat. It needn't be taken to a winch; just snug it by hand on the cleat to secure the boom against inadvertent jibes.

Whenever a problem is found with the standing rigging, the first measure is to take all pressure off of that component. Tack to remove loads on a shroud, head up to protect a backstay, and sail downwind should the forestay be threatened. Next, directly relieve the pull by attaching a halyard to the stem, transom, or chainplate.

Tighten the halyard to approximate the normal tension so that the mast keeps its alignment. Once the rig is protected, take steps to make repairs of the broken component.

I recently outfitted *Voyager* with a Technora boom topping lift strong enough to replace a shroud or stay that failed. This synthetic fiber is a braided line, with an inner core surrounded by an outer jacket. These fibers are very often braided or plaited, as opposed to three-stranded nylon, and lie within a protective cover. This outer covering decreases exposure to chafing and sunlight, while all laid rope is subject to such degradations. The covers are given different textures and colors to denote specific purposes and for improved grip.

Stronger than like-sized stainless steel, Technora combines incredible strength with light weight and low stretch. It's also heat stabilized to minimize shrinking and hardening of the rope fibers. It serves as a perfect backup system for failed running rigging, so I no longer carry a long length of 1:19 stainless steel on board.

Even though the lower terminals and their attachments to chain plates are clearly visible to the naked eye, a majority of the standing rigging is located high enough from the deck to make inspection difficult. Use binoculars to examine all upper connections of shrouds/stays to the spar. If you find anything abnormal, take a trip aloft in the bosun's chair, if conditions permit.

As mentioned, lower swage fittings are at more risk than those higher up. A cracked or corroded terminal swage fitting can't be replaced at sea because most sailboats aren't equipped with swage machines, so we use alternative methods with other fittings. The wire must be cut just above the defective fitting. Large bolt cutters work best, nipping away at wire strands until they're all cut. Don't spend time trying to use a hacksaw; it'll be time wasted unless you can hold the wire securely in a vise.

Once the wire is cut, a time-tested method of repair is to place an eye with thimble onto the end using cable clamps (Fig. 5.25).

*Fig. 5.25
Cable clamps are used to fashion an eye into the end of a wire.*

I carried several such clamps as standard equipment on all offshore passages for a lot of years, and fortunately never had to use them.

Cable clamps consist of two parts: the saddle and the U-bolt. The saddle lies against the standing portion of wire after it has formed the loop around a thimble. The U-bolt then fits into openings in the saddle and is tightened to secure the loop. Be sure to use three cable clamps and tighten them securely; the clamps each face with the *curved part* of the U-bolt against the *standing part* of the shroud (Fig. 5.26 below).

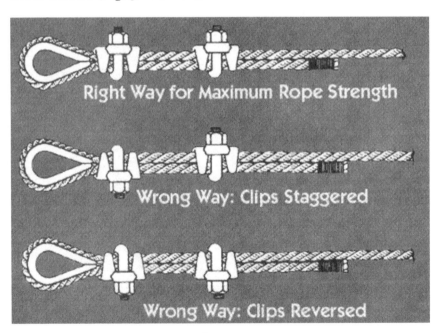

*Fig. 5.26
This demonstrates the correct placement of cable clamps for maximum effectiveness.*

In addition to cable clamps, a six-foot piece of wire the same size as the rigging on the boat can be fitted with a stud on one end (sized to fit the boat's turnbuckles) and an eye or fork on the other (to emulate the rigging's fittings aloft). In the event of a break at the deck-level fitting, the stud can be fitted to the turnbuckle and the short cable then clamped to the remainder of the broken stay. The same holds true for the swage aloft. If the wire breaks in the center, the six-foot piece can be fixed to both broken pieces of the stay.

Now we have more modern fittings that we can attach ourselves while at sea. These aren't used to form eyes at the ends of wires; they're actually replacement fittings. The Norseman, Sta-Lok, and CastLok systems all separate the outer individual strands of wire from the wire's inner core. A socket slides over the wire end, with threads facing the end. An inner component is then placed at the end of the inner core of the wire, with the outer wires overlapping. A terminal component, with the eye manufactured-in and threads to match the socket, slides over the inner cone and screws into the socket. When these are screwed together tightly, the terminal squeezes the outer wires down onto the inner cone to form a strong end fitting. These systems have an advantage; they may be disassembled and inspected to ensure a strong terminal fitting.

Once the cable clamps have been used to form an eye, or a new fitting has been installed, we bridge the gap between shroud and chainplate with a series of shackles, shackles with chain, or loops of Technora line to complete the repair. A new shroud (or stay) complete with a swaged fitting should be fitted by a rigger when the boat makes landfall.

Broken end fittings at the spar end are more difficult to repair because they can be integral with segments within the spar (Refer to Fig 5-2). These fittings can wear out or corrode with very little outward evidence. The only way to be sure they're solid is to remove the shroud fitting and inspect them directly.

Loss of a mast receptacle unit is very difficult to repair. It requires a method of attaching the stay/shroud to something that replaces the broken component. This also entails considerable time up the mast in a bosun's chair, which is usually difficult and often impossible at sea. Immediate boat maneuvering to relieve strain from the affected section is mandatory. Supporting the mast with a halyard (or Technora topping lift) is the second step.

The most practical strategy is to lash a line around the spar at the point of breakage to serve as an anchoring point. This is very effectively done with an icicle hitch, which is excellent for connecting to a post when weight is applied parallel to the post. This knot will stay in place even when holding a substantial load on a smooth surface, even to a tapered post, hence the name "icicle hitch."

Tying the icicle hitch:

1. With the spar in front of you vertically, begin making loops around the pole with the line's working end, extending upward away from the direction of pull. The number of turns depends on the line friction and the amount of force on the hitch; use at least five turns for this application.

2. Tighten the wraps around the pole so they lie together neatly with no gaps or overlapping. Bring the working end back over the wraps and place it loosely across the pole, and then twist the line to create a suspended bight.

3. Bring both the working and standing ends up and through the bight thus created, and then extend the bight around the end of the spar.

4. Pull both ends of the rope taut, and remove all slack from the completed hitch.

5. The two ends of line are now available to connect to shackles, other lines, etc., to reinforce the spar.

For video instruction on icicle hitches and other nautical ties, visit these web sites:

www.expertvillage.com   www.marinews.com

In our example of a lost shroud, the hitch would be placed on the mast to approximate the position of the shroud attachment. This would also be the knot of choice for loss of the mast; tied to the stump left over after the mast broke, or to the top of another spar (a spinnaker pole or the boom) used to jury rig a replacement.

Other methods of making an attachment to the broken spar include drilling a hole through the spar for a point of attachment and attaching a line to halyards/cables protruding from the end of the mast. In our example of the lost shroud, once the point of connection is established, attach a stout block to this line, and then loop another line through the block with both ends extended toward the deck. Secure one end to the toe rail to approximate the support of the missing shroud, then take the bitter end to a winch, haul it in tight, and secure this to a cleat or line clutch.

If the deck end of a shroud terminal fails, we can re-use the same shroud. Place a new terminal on the wire end, as described, or fashion an eye using cable clamps. Secure the shroud with a series of shackles, or chain and shackles.

Loss of a shroud or stay can cost you a mast. Should a stay break, take the same immediate action of maneuvering the boat to relieve all tension from that direction. Next, fasten and haul in a halyard or Technora topping lift to approximate the position of the stay. Assuming that the mast is still standing and a halyard is now taking the load, we're faced with the same situation as with loss of a shroud--except the mast attachment is much higher. Our jury rig is easier in one sense, though, because we've got the masthead fitting very close to our attachment point. We can attach a block to a convenient position on the masthead, either directly or with shackles. The parted stay can also be re-used with methods similar to those used on shrouds.

Thus far we've considered methods of standing and running rig repair while under way. Now we'll shift the focus to what happens when the mast does come down.

Fig. 5.27
Fractured main mast on a ketch.
What appears as a baby stay kept the lower main mast up after the primary forestay was lost. The upper spar toppled aft after the fracture occurred at the point of greatest deflection.

Masts usually break because a supporting structure fails or the boat rolls over. In the case of the former, should a forestay or backstay part, the mast is probably coming down unless some very fast maneuvering takes place. Breaking a spreader or shroud *may* give more time to relieve the tension on the affected side before losing the rig.

Dismasting can be a sudden and traumatic event that changes the tenor of a passage abruptly, from sailing along to one of survival mode. The loud snap of the rigging is accompanied by a stiff jolt felt throughout the boat. That, followed by the resounding thump of a spar and its rigging against the decks can be terrifying. The boat loses way and often yaws sideways to the seas, with waves breaking on the beam. A new motion takes over as the center of gravity is lowered: the vessel rolls from side to side in quick, deep arcs that make footing treacherous. Capsize or rollovers can cause the most damage, and greater chances of crew injury or death as well (Fig. 5.28, below).

Courtesy of U.S. Coast Guard

Fig. 5.28
***Forestay failure on this trimaran left the vessel
without its rig lying helplessly broadside to the waves.
The stricken crew awaits rescue.***

We had done very well in the opening day races of the SORC Championships in 1997. Our Hobie 33, *Stealth*, was fast and we had a good crew with high expectations. During pre-race maneuvering on the second day, I was paying attention to the moderate winds, getting a feel for the timing and magnitude of shifts. The other crewmen were all busily occupied as usual; it was just another day on the race course.

The snap of the mast was surprisingly loud in the warm morning air; and then the boat lurched and rolled to leeward. We looked up in disbelief as the mast began to topple and then fell heavily on the starboard side. It hit the deck with a sharp thud, having just missed the crew in the starboard cockpit. Just like that, our day and regatta were over.

The skipper must come to the fore in this situation and display calm leadership to a confused and frightened crew on a boat suddenly plunged into disarray. It's worse yet if this all happens in the dark of night.

There are immediately two concerns: the health of the crew and securing the boat against further damage or being broadside to large waves. Take a head count to make sure nobody has been knocked overboard, seriously hurt, or in need of life-saving first aid. If a crewperson requires immediate care, the most qualified person assumes that responsibility. Search out every member of the crew to assess injuries. Tending to any serious injury takes precedence over the boat, unless she's in imminent risk of sinking. Injuries range from minor abrasions and contusions to head trauma, fractures, shock, and near-drowning. Immediate first aid measures may include applying pressure to stop bleeding, bringing the victim of head injury to a safe place to rest and be examined, initiating CPR to a near-drowning victim, and stabilizing a broken limb. If possible, contact land bases for emergency medical advice and begin rescue operations. Please refer to **Further Offshore** for a comprehensive discussion of the medical kit and treatment of offshore illness and injuries.

In the meantime, another person must tend to the boat. Shut the engine off immediately if motor-sailing. Don't risk making the situation worse by having a line wrapped in the propeller. Is the mast still on deck or is it overboard? If it has gone over, try to maneuver the boat so that the spar and associated debris are to the leeward side, away from waves that can dash them against the hull and cause further damage. Next, steer the boat toward the waves to reduce deep roll motion. Consider deploying an anchor or sea anchor to keep the bow into the wind and waves.

If sinking is a possibility, prepare to deploy the life raft (see Chapter 11, Abandon Ship) over the side to leeward, and ready the abandon ship bag, handheld GPS, satellite phone with portable antenna, EPIRBs, and other items selected for the raft. If possible, send a message by satellite phone or handheld VHF radio--Single Sideband won't work if the antenna was incorporated into the backstay--giving your identification, position and condition of the boat and crew.

The crew should don life jackets/survival suits and harnesses and take seasickness medication immediately after a dismasting, and be very careful maneuvering about the boat. Lines, shrouds, and perhaps the mast and boom are strewn about the decks, the seas are often up, the decks are wet, and footing is treacherous.

Once the crew has received the attention they need, the vessel damage must be assessed and plans formulated. All able crewmembers should participate in the aftermath assessment, delegated tasks by the skipper. Examine the boat for leaks and fire. The batteries would be shorted out and ruined if flooded in a roll-over; loss of battery power alters the situation even more. If the batteries aren't flooded, move them to higher ground if necessary. Check the water supply for seawater contamination. Find out if the engine runs, and check the fuel. Be certain that no lines or cables threaten the propeller before engaging the gears.

Keel-stepped masts are supported more securely than their deck-stepped counterparts. They often survive capsizes and sometimes even roll-overs. When they break, they're more likely to snap in the vicinity of the first spreaders. This is an important factor when considering mast options for offshore vessels. Masts stepped on deck with supports below are more readily lost in these situations, parting right at deck level, leaving gaping holes in the deck.

Though still connected to the boat, dismasted spars are generally swept overboard to the lee side if the sails were full. If the boat rolls through 360 degrees, though, the mast ends up toward what was windward. Quick action may be necessary to salvage the spar before it sinks or does further damage to the hull, especially if it's situated to windward. If the spar threatens to hole the boat, the best action may be to cut it loose. Salvage is the best option, though, if possible. Use a boat hook to bring the spar in close and lash it to the hull, using fenders as cushions.

Once the mast is secured against sinking and poses no further danger to the hull, make preparations to bring it back on deck. The first step is to free the spar from its deck connections. Try to unscrew the turnbuckles, toggles, and other fittings rather than automatically cutting them with bolt cutters. Once the lower fittings are removed, leave the upper attachments on the mast if at all possible. Only cut the shrouds if absolutely necessary; the more you cut away now, the more rigging it'll take later. Take the boom off, along with radar and masthead appliances, to make the mast more manageable.

With the mast lashed to the hull, its stays and shrouds disconnected from the deck and appliances removed, now comes the task of bringing it up over the gunwales and lifelines. Do not risk a crew-overboard situation or injuries by attempting to raise the spar in rough conditions. As long as it's not harming the hull, it's advisable to wait for waves to diminish.

One way to make lifting easier is by rigging a gin pole (Fig. 5.29, left) by using the spinnaker pole, remaining mast stump, a whisker pole,

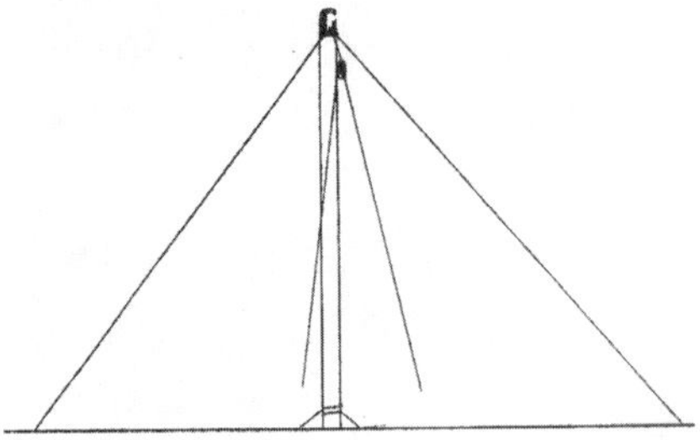

Fig. 5.29
A gin pole is a temporary structure used to position a block and tackle mechanism high enough to elevate a spar into its correct position. It is supported fore, aft, and side to side with guys, and also at its base. Here, a spinnaker pole has been converted into a makeshift gin pole.

RIGGING FAILURES

or a mizzen mast. The gin pole can be useful in retrieving a spar onto a boat or to hoist the spar into position on deck.

The gin pole utilizes mechanical advantage to make the work easier, but the mast can still be difficult to control when hoisted from only its midpoint, and rolling may prohibit erecting a gin pole or climbing the mast stump. The other option is to attach several lines to the hull, extending them over the lifelines (if they're still up) and underneath the mast, and then back again (Fig. 5.30, below). When these lines are pulled in unison, the mast can be elevated as a unit.

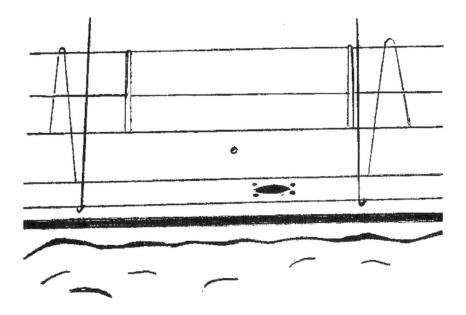

Fig. 5.30
This may be the easiest and safest method of mast retrieval because it doesn't involve erecting a structure on deck and brings the mast up under greater control. Here, the spar is elevated with lines over the lifelines and back to the deck.

Crewmembers should always wear life jackets and harnesses if working in rough seas. Carefully loosen the lashings that hold the spar in place, and station one crewmember at each line. The mast will be elevated, maneuvered over the lifelines, and lowered to the deck more safely when hauling the lines in together.

With the mast and its gear finally on board, and the transom assuredly cleared of all debris and lines, you could start the engine and begin making progress to the nearest port. If there's any question about the propeller, place the transmission in neutral and turn the driveshaft by hand, or have the strongest swimmer, attached with a line, dive overboard for a look if the seas allow. Do not engage the propeller without being certain it's safe.

If the engine is unusable, remain safely on the sea anchor while the real jury rigging begins. There are some principles that apply to boats in general, but how to jury the rig will depend on the circumstances. If a mast stub remains, it's probably broken just above the lower shrouds, either above or below the first spreaders. This is a great benefit, because three possible strategies could be used:

Use lashings (icicle hitch) at the upper end of the stub, or connect a shackle directly with screws (it's nice to have a cordless drill about now).

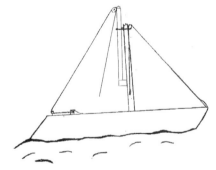

Attach blocks to the outboard boom end, and fold the boom upward against the mast stub to serve as a spar, as in Fig. 5.31.

Hoist the spinnaker pole to the upper limits of the track to function as a mast.

Fig. 5.31
The boom with a block on the outboard end has been elevated against the spar stub, lashed into position, and will now serve as the jury mast.

If conditions permit, scale the stub and secure lashings to support a block. A line through the block can hoist the corner of a sail aloft. It may even be possible to drill and tap screw holes to mount a block directly to the spar.

The boom can be elevated from the aft side of the mast stub, while the spinnaker pole can be extended vertically in the same way from before the mast stub. Once hoisted into vertical position, either the boom or spinnaker pole is lashed to the mast, where it then serves as a makeshift mast. This is a reason to have a spinnaker pole on board, even if you don't often fly symmetrical spinnakers.

If the mast stump is shorter, a section of mast can be inserted into the stump and secured with bolts all around. The structure, guyed on four sides, becomes the jury rig.

Total dismasting poses the most difficult situation. With no stump remaining, the replacement has to be hoisted, held at the base, and stabilized with guys to make it useable.

First determine a method of capturing the mast heel. The best way is to get it below the decks onto the sole or keel. This may entail removing the mast partners, installing a support post below, or cutting a hole through the deck. The base could be bolted to the sole, lashed very securely, and so forth. This provides the most solid jury rig in this situation. Depending on how much the mast weighs and how easy it is to control, you may end up cutting it to a more manageable length before attempting the hoist.

Attach lines (which are easier on the hands) to the shrouds, and designate crew to control them. The lines can be tied rapidly to the toe rail when the mast is vertical. Use halyards or tie lines to the forestay and backstays, and coordinate those controls as well. If the spar is to be directed through a deck opening, tie shorter lines near the mast base to be used in controlling its descent below decks.

Methods to hoist the mast upward, toward vertical, and to control the heel must utilize mechanical advantage. Here are three possible methods:

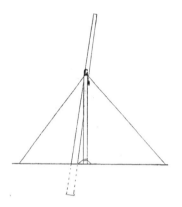

Fig. 5.32
Hoisting the spar with a gin pole. The spar is guided by its shrouds and stays while being gradually elevated. Another line from below, used to maneuver the mast downward toward the keel, is lashed to the spar just above its base.

1) Employ a gin pole, hoisting from a block near the top. (Fig 5.32, above).

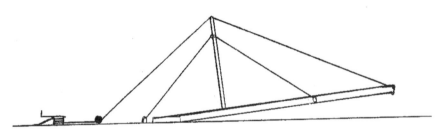

Fig. 5.33
The boom, stabilized in position with guys perpendicular to the mast, provides the elevation needed to hoist the spar to an upright position. The main halyard is led through a boom end toggle sheave and then taken to a block on deck. The halyard leads to a deck winch for hoisting. The shrouds and stays would be connected to the mast, with lines tied to them for control as the spar is elevated.

2) Use the boom as a lever, hauling on a halyard or topping lift to hoist the spar (Fig. 5.33, above).

Rigging Failures

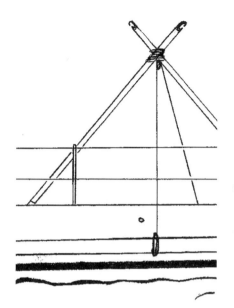

*Fig. 5.34
A supporting structure
is assembled from two
spinnaker poles in this
example. This structure
requires guys fore, aft, and
laterally, and both legs are
captured and stabilized to
provide a solid base. A line
through the block is hauled to
raise a spar onto the boat, as
pictured here, to guide it into
its deck position, or to serve
as a spar itself.*

3) Construct a support structure using spinnaker pole, whisker pole, or boom. Two or three poles are lashed together near their tops to form a hoist with a block (Fig. 5.34, above).

Whatever strategy is employed, this operation can go surprisingly well if all lines are controlled and the seas cooperate, but it can also be difficult and dangerous when hoisting a heavy, unwieldy mast in the wrong conditions. All guy lines must be tended with continual adjustments as the spar is gradually elevated. Those involved in the hoist should proceed slowly, watching as it goes up and staying alert to shrouds that snag or guys that need adjustment. Others will be needed to guide the heel into position, and to tend lines to control the descent. When everyone is ready, begin hoisting as a unit with the chosen mechanism. Keep the guys tight, yet ready to adjust as the spar changes position.

If the mast is being returned to its original assembly, the base must be stabilized. If the intent is to rest it on deck, multiple lashings will be needed to keep it in place as the spar is elevated. Placing a section of spar into a mast stub can be the most difficult maneuver, because it entails raising the whole rig, base and all, to a higher position. Once high enough, the base is directed and maintained at the top of the stump, while the opposite end is hoisted further. Once the spar is high enough, the heel will begin to enter the stump, and it must be controlled to prevent its slamming downward. If the goal is to lower it through a deck opening, the heel will have to be stabilized in position with multiple lines. As the spar is elevated, it must be directed toward and through the deck opening, and then its downward descent must be controlled.

When the mast base reaches the keel, it's bolted and/or lashed into position. Shrouds and stays are fixed into place as tightly as possible. This may involve returning the original hardware into position, or affixing lines to the toe rail.

Once the spar is hoisted and secured into a workable position, go about the business of configuring sails. Take advantage of sail lift if at all possible by deploying a foresail that directs air past a mainsail. Lift is gained as air passes over both sides of the foresail, and even more lift is generated when the air is directed across the main. Set the normal sails if the entire mast has been repositioned or if a jury rig provides near normal height of the mast.

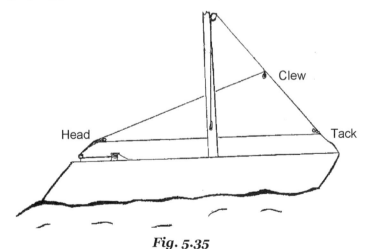

Fig. 5.35
This jib has been hoisted by the clew on a shortened spar.

Sails don't have to be hoisted in their normal configurations (Fig. 5.35 opposite). The object is to get as much sail material as possible exposed to the wind; hoist sails in whatever position works most effectively.

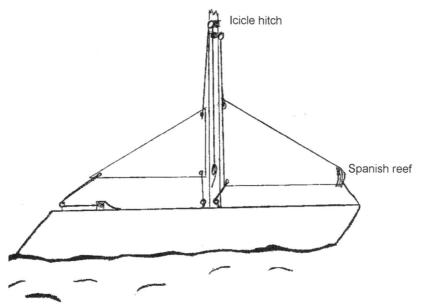

Fig. 5.36
This rig uses two jibs to create lift in a workable sail plan. Note the use of the Spanish reef and icicle hitch. Smaller jibs are easier to handle, and can split the jury rigged plan into fore and aft sails.

If the jury mast is shorter, it's still possible to set the regular sails, but the tops will have to be shortened by using a Spanish Reef (Fig. 5.36, above). This is done by tying a short, stout length of line around the sail, beneath the head. This line is attached to the halyard to hoist the shorter sail. Connect the halyard to the clew. What was the head now serves as a clew, and the tack remains the same. The mainsail or a suitable jib can be hoisted in this manner and tacked fore and aft to create a "sail plan" (Fig. 5.36).

Be prepared to think abstractly or "out of the box" in order to create usable configurations to suit the circumstances and make the best use of the materials on hand. There's no rule stating that sheets have to lead to their normal tracks or traveler; a spinnaker pole can become a mast, and a sail can be deployed upside-down if that works best. Attach blocks to the toe rail, stem, transom, or whatever position creates the most advantageous set of the sails. The conformations vary depending on what sails are available and the ingenuity of the rigger.

6

Loss of Steering Control

The midnight to 0400 watch is now into its second hour. A lone crewmember is on duty, while the rest of the crew is grabbing some much-needed rest after three days of Force 9 winds and waves more than twenty feet high. Conditions have begun to moderate now, and the latest weather download showed the low moving away to the northeast. The rest of the passage should get easier; even a warm dinner could be in store.

He settles down into his favorite spot in the cockpit, nestling up close to the dodger where the spray can't reach. The warm coffee tastes good and helps ward off the chill of a night at sea. He's watching the phosphorescence glimmer on the wave tops when suddenly, without warning, a wave smashes against the starboard quarter and the boat veers wildly to windward. In an instant the cockpit is deluged with seawater as a second wave--now from the port bow--breaks over the boat.

Scrambling out of the comfortable nook, the confused crewman dashes to the pedestal; the autopilot is still on but now sounds the Off Course alarm. He grabs the wheel to bring the boat back onto course, but is startled to learn that the rudder will not respond; he attempts to come to grips with the awful realization that he's lost steering control. Now alarmed *and* confused, he gazes up at the sails that are luckily illuminated by a nearly-full moon, and the reality dawns on him at last: the helm won't respond because the boat is now hove-to on the other board!

The sinking feeling resulting from even this short-term loss of rudder control was palpable as I--yes, I was the crewman--contemplated the alternative methods to steer the vessel in the midst of those tossing seas.

Of all the systems on a boat, steering probably gets the least attention, maintenance, and thought to coping with its failure. Many sailors never check the steering gear, let alone go over it carefully at least yearly. Part of the reason for this is that today's steering gear is so good; from the wheels to the rudder, the equipment is of such a high quality that it usually works flawlessly for years, especially in the relatively light duty of weekend outings or day sailing. What would happen if your boat were subjected to a week-long passage, especially when the winds piped up and waves built to 20-plus feet?

Where's the emergency tiller? Does it affect the compass deviation? What steering system is used inside the pedestal? Are the cables all tight? What about the sheaves, and the connections to the radial drive? And how about the biggest question of all: what's your plan for steering the boat if your rudder drops out of her?

These are questions every sailor should be able to answer, and surprisingly few can. If this chapter does one thing, I hope it sends the message to pay attention to this important component, because losing it is real trouble.

I discussed assessment of the steering at length in *Ready to Sail* with this same goal in mind. It shouldn't be that only surveyors, delivery captains, and boat yard employees delve into the aft cockpit locker to examine the steering system. Maintenance of this equipment is not difficult, and periodic examination of the system helps you to know its workings, which is invaluable if something breaks or malfunctions. I encourage all sailors to watch their heads as they crawl into this mysterious realm, and get familiar with the landscape of the steering gear.

By far the most common system in use today is chain/cable steering utilizing a chain from a sprocket on the wheel shaft. Both ends of the chain extend down through the pedestal, where they connect to fittings attached to cables (Fig. 6.1 below). The cables lead to sheaves near the pedestal base, and turn toward a series of sheaves that direct them toward the quadrant or radial drive unit connected to the rudderpost.

Fig. 6.1
A steering cable, from its attachment to one end of the chain, extends down through the pedestal toward a sheave near the base. Notice that the cable, because of its acute turn around the thimble, has become defective.

The cables encircle the quadrant or radial unit, where they usually terminate in an eye with cable clamps bolted to the assembly. (Fig. 6.2, page 102)

Fig. 6.2
The rudder tube (encasing the rudder post) is shown as it extends downward from the steering quadrant. Steering cables are seen leading from the pedestal base (to the left) and around the quadrant, where they're terminated with cable clamps.

From the rudderhead fitting, the rudder post extends downward toward the rudder tube, through which it exits the hull. The rudderhead is extremely prone to corrosion that leads to weakening and possible failure as water leaks downward past the upper bearing.

Obviously, there must be a method in place to affix the rudder post, keeping it attached to the boat. These methods vary widely, but often involve a fitting or flange atop the rudder tube where a pin goes through the post. Sometimes the post is attached closer to the rudderhead fitting. Trace your rudderpost as it emerges and extends upward from the rudder tube. Look for its mechanism of attachment; what keeps the rudderpost in the boat. If the vessel has two rudders, check the post that connects the individual rudderposts.

In a vessel with hydraulic steering, you'll see copper pipes extending from the pedestal. These should connect to a bypass valve and then extend to the hydraulic piston assembly. A tiller arm extends from the piston assembly to another arm on the rudderpost, where it attaches to turn the rudder. Most problems with hydraulic steering involve leakage of the hydraulic fluid. This is usually a red, viscous liquid found near connections in the hydraulic system. Fluid loss results in weakened responses, or failure.

Electronic autopilots control rudder motion with a drive unit that either sends an armature to the quadrant, or uses a gearing action to turn the quadrant or rudderpost. Fig. 6.3 demonstrates a quadrant with autopilot drive and sensor arms attached.

Fig. 6.3
The autopilot drive and sensor arms are seen attached to this steering quadrant. Metal shavings indicate an impending problem as the cable is being damaged.

These systems can be electronic or they can use hydraulics. Check the mechanism for broken or slipped gears, or debris within the gears, objects (*e.g.* fenders) preventing quadrant movement, disengaged wiring, etc., that interrupts autopilot function.

Rudderstops are a key component of any steering system, and can be the site of steering difficulty. Steering failure can occur when the rudderhead fitting is slammed violently against the stopper and becomes wedged into a fixed position at the furthest extent of the steering range. The rudder will be locked into a hard turn, either to port or starboard, and no manipulations of the wheel or tiller will change course. For this reason, the wheel must be controlled at all times, especially against following seas or when backing the vessel at even moderate speeds. Whenever the boat is at anchor or dock, the wheel should be centered and locked into position to stop rudder movement if waves approach from astern and turn the rudder hard over.

Rudderstops consist of a fitting, or arm, usually seen on the rudderhead or rudderpost, but also on a quadrant or radial drive unit. The rudderstop arm extends directly aft. As the rudder turns from side to side, the fitting meets a shock-absorbent stopper to control the rudder movement. This controls the rudder's excursion and limits the extent of the steering arc. There are stoppers at the port and starboard extents of the steering range.

Rudderstops have a variety of configurations, but all should be strong, unable to wedge or jam into position, and ideally mounted independently of the steering mechanism--i.e. positioned on the rudderpost, not the quadrant or radial drive unit.

When the rudderstop jams, it's often because the stopper is not in the right position to engage the fitting squarely, or may not be large enough to prevent the rudderstop from becoming wedged against or underneath it. Perhaps the rudderstop is not mounted securely enough on the rudder post or drive unit, and has become dislodged or broken off. Repairs consist of dislodging the arm from its locked position, and the fix depends on the individual components involved.

When steering trouble arises, determine what area of the system is involved. Listen for abnormal noises, such as grinding of metal. Feel the helm while steering; look for grating, difficulty in moving the wheel, no resistance to turning the wheel, or a locked helm that won't respond. If problems seem located between the pedestal and rudderhead, the segment where most problems arise, begin to troubleshoot the chain/cable or worm gear systems at the quadrant or radial drive unit, where the cables are easily viewed. Observe the cable terminals as they connect with the drive unit for kinks, laxity, chafe, or corrosion. Metal shavings are always abnormal, and often signify cable chafe or damaged sheaves. The cables must run fairly, without contacting any ship's components or each other. Any damaged cables should be replaced.

Cables usually change direction at least once in their run from the pedestal, and sheaves can be the source of problems. If they loosen, cables can jump off the sheaves or, if corroded, they can seize and prevent smooth motion. The cable gradually deteriorates as it moves over the fixed sheave.

Now observe the rudderpost as a mate turns the wheel. If the post is locked, examine the rudderstops carefully, especially if the helm is locked over and won't move. Check the rudderhead and quadrant/radial drive unit area for debris jammed into the mechanism, preventing motion. Offshore, this can easily occur when gear or garbage stored in an aft cockpit locker shifts position and becomes lodged within segments of the steering gear. Whenever possible, especially at haul-out, check the rudderpost connection to the rudder. The post enters the upper rudder surface and is connected to a stainless steel web structure around which the fiberglass rudder is fabricated (Fig. 6.4, above).

Fig. 6.4
The internal stainless steel framework of a rudder, with outer shell removed.

Courtesy of Scanmar International

The internal web/fiberglass rudder is rigidly fixed into relative position; the rudder should not move around the web or rudder post. The web can deteriorate if water enters the rudder and instigates corrosion of the metal. The area just above and below the top of the rudder blade, called the lap zone, is where most inner web corrosion begins. The corrosion is worst just inside the rudder, often where the post is welded to the internal webbing system. Over time, the rudderpost's stainless tube or shaft can corrode to the point of total failure and rudder loss when great torque is placed on the rudderpost.

Stray electrical current can also have its galvanic effects on rudderposts. If corrosion of the propeller or the shaft zinc is evident, close examination of the post is warranted. This can be controlled by installation of a galvanic isolator, bonded to the rudderpost and/or stock. The wisest course of monitoring the lap zone is to periodically grind away the fiberglass atop the rudder blade to expose the post in that area, then refill with epoxy to seal it.

Water ingress also leads to delamination of the fiberglass rudder structure. Most rudders are built of cored fiberglass and are prone to delamination and physical damage, just like fiberglass hulls. Since fiberglass and stainless steel expand and contract at different rates, keeping that steel/fiberglass bond water-tight is difficult.

If the rudder/rudderpost connection is compromised, the rudderpost will not turn the rudder effectively, if at all. If the rudder seems sluggish, i.e the boat responds very slowly or not at all to wheel movement, get onto the transom or the water and observe the rudder while the wheel is turned. If the rudder/web mechanism parts, and the rudder loosens from the post, it won't drop out of place immediately because buoyancy maintains the rudder within the rudder tube. If a wave lifts the transom, however, and takes the rudder out of the water, it would then be free to fall away from the hull.

If the cables are connected to the quadrant/radial drive unit securely, if they run fairly past all sheaves and enter the pedestal cleanly, if the rudderstops are clear and the post checks out, troubleshooting proceeds to the pedestal.

Before opening all the access panels to the pedestal, turn the steering wheel; it should elicit motion of the shaft in the pedestal, whatever system is used. Any failure here—such as hesitancy, difficulty, or a lack of turning—is a signal that the problem is most likely at the forward end of the steering system, within the pedestal.

You'll have to open at least one panel and perhaps even remove the compass to gain access. Chain/cable systems begin at the steering wheel, where the wheel shaft inside the pedestal holds the sprocket. The sprocket in this system (and gears in alternative systems) is fastened to the shaft with pins and sometimes a locking key. Check these pins and keys for wear or shearing. You should also confirm that the gears mesh properly and that they are adequately greased, with no stripped teeth. It may be necessary to replace some components (worn gears or pins, a sheared key, or ball bearings) if they're damaged or corroded.

The wheel is normally held onto the shaft with one or two screws, and turns on bushings or ball bearings. Observe these while turning the wheel. Next, examine the sprocket for bent or missing teeth, making sure the chain is clean and well lubricated. Steering chains are rugged and rarely fail, but they can come off the wheel sprocket, become corroded, or disconnect from the cables. Check the cables in the binnacle for frayed wire or defective terminals. The cables terminate in an eye/thimble and any time a cable is wrapped severely this way, the strands are placed under strain and failures are apt to occur, as in Fig. 6.1.

In our vessel that lost its steering control, one of my considerations was that the rudder had fallen off or was broken.

This is one of the most-dreaded of scenarios for sailors. How would you know if the rudder was lost? Either the rudder disengages from the inner web structure and drops away from the rudderpost, or the post comes loose from its connection within the boat. If the post and rudder both drop off, there's a good chance that water is entering the hull, maybe very rapidly, through the rudder tube. One fast look into the aft locker will tell. Loss of the rudder only, while the post remains, would at least keep the rudder tube patent—as long as the post was solid. Hopefully, proper maintenance alleviates the chances of either actually happening.

Now, what to do when the steering fails, and the boat is at the mercy of the elements? It depends on what segment of the steering is at fault. If the problem lies before the rudder post, there are two methods of steering to choose from. You can bring out the emergency tiller, set it into its slot atop the rudderpost, and drive with it while servicing the problem. This is a maneuver that all crews should have practiced; know where the tiller is and how to deploy it. Some tiller arms are actually too long, and the steering wheel has to be removed for it to work.

The other method for steering in this situation: don't forget the autopilot. It should still drive the boat, since it connects to the quadrant or rudderpost independently of the forward steering gear. An auxiliary rudder-type wind vane, though not designed for the purpose, can also replace the function of the primary rudder. These units are large enough, are mounted to the boat very solidly, and can at least temporarily take on that load.

If these methods can't work, I would heave-to, set the sea anchor or the iron anchor, and set about sorting out the problem.

Examine the components according to where the problem seems to be. With knowledge of the gear, troubleshooting should proceed logically until the difficulty is located. Repairs could involve returning a cable to its sheath, replacing a retaining pin that holds a cable to its chain end, securing a cable to its quadrant, or tightening a loose cable. Perhaps a hydraulic system has leaked and needs fluid replacement, or a mop handle has jammed into a quadrant.

Part of voyage preparation is ensuring that you've got spare parts or pieces that may be necessary to repair or jury rig the steering mechanism. Electronic autopilots are sectioned into distinct component parts: the display, fluxgate compass, central processing unit (CPU), and drive unit. These are essentially enclosed and not amenable to "McGuivering" at sea, but there are things to check when they malfunction. The most common problem is to have a metal object placed too near the compass. This electromagnetic interference disrupts the compass, and the boat fails to hold a course and may prefer driving in circles. Make sure the crew understands to keep the fluxgate clear and not store anything metal nearby. I had trouble once when a gallon container of teak oil was moved to a locker close to the compass. Simply moving the can got us back on course.

The system is connected to 12-volt power, with its own circuit breaker and an on-off toggle switch. The power must be on at the control panel, and the toggle in the On position. Many electrical connections are made throughout the pathways, and all are subject to corrosion and physical disruption. If the display fails to turn on, troubleshoot the electrical input from the control panel to the display unit. Behind the display, there are connections for power and for integration of the display with the CPU and other functions, such as the speed, wind, and depth sounder. Check these connections by removing them, and then sand or lightly scrape off any corrosion and replace very securely.

The CPU is a complex network of connections that all have to transmit effectively. I've returned autopilot function by simply disconnecting wires and replacing them firmly. Check the unit's fuses; a blown fuse isn't good. There should be a spare fuse attached directly to the unit. Make sure you've got fuse replacements.

At times, you can bring a system back by turning off the toggle switch power and then turning it back on, in effect re-booting the mechanism.

Now move on to check the drive unit. Wires from the CPU must enter it cleanly without chafe along the pathway. The unit has to be *very* solidly mounted against a bulkhead. Observe the drive armature as it spans the distance to its target (usually the quadrant or radial drive unit) and connects there with a pin and cotter pin. A sensing arm leaves the same area, returning to the drive unit. You'll see a message at the display unit reading "Drive Stop" if something prohibits the armature from controlling the helm. This is often a fender or some other gear that falls into the wrong place and won't allow the quadrant to move. I've also gotten that message without any obstructions, and solved the problem at the CPU by tinkering with wires (disconnecting, cleaning, and then reconnecting).

A disconnected sensor arm may cause intermittent or total loss of steering control, resulting in an alarm from the display. Suspect this if the problem occurs after sailing in heavy weather with a lot of rough motion. In this situation, the fix may just be a simple matter of replacing the cupped aperture to its ball attachment at either end.

Aside from confirming electrical input, and that all components are connected securely without corrosion or interference, individual components of autopilots must be replaced when they fail at sea.

Wind vane systems translate comparatively weak signals from the air vane through the servo mechanism that amplifies the force to enable course correction. The main components are the air vane, the intermediate motion transfer mechanism, a paddle in the water, and lines connecting to the steering wheel.

The best systems supply two air vanes: the standard unit and one much larger for light air conditions. The vanes must be installed with the correct surface facing the wind. Configured improperly, steering reactions will be reversed, taking the boat farther off course instead of issuing a correction.

To protect the vane against damage, one recommendation is to remove the wind vane from the vane gear when not in use. Eventually there will be some wear of the vane and its mount if it's left banging from end stop to end stop for extended periods of time. Another tip: secure a lanyard to prevent loss of the vane should it work loose of its attachment. Vibrations from the wind or engine can loosen attachments; use Loctite to prevent disengagement of connections.

Proper installation is a crucial element in a wind vane's performance. It should be mounted exactly according to manufacturer's recommendations, on the centerline of the stern whenever possible, measured both vertically and horizontally without tilting forward, aft, or laterally. Servo-pendulum units require no special measures to strengthen the transom, and are usually mounted with only the stainless steel washers provided. Auxiliary rudder type steering systems generate much higher loads and definitely do call for additional reinforcement. Mounting positions of all units at the corners of the transom, where the hull is strongest, is highly recommended with either design.

The servo-paddle is vulnerable to damage from collision, whether it be with objects in the water, other vessels, or docks. The boat should be docked with the transom inward, and crewmembers must be ready to fend other vessels in crowded settings. In periods of down time, remove the paddle from the gear entirely.

The paddle should be attached to a line that acts as a safety lanyard in case of collision; this arrangement is also useful to pull the paddle out of the water for storage while sailing with the vane disengaged.

Steering corrections from servo-pendulum wind vanes (Fig. 6.5, following page) are initiated when the air vane is moved by altered apparent wind direction. This motion is relayed to a servo-paddle in the water that is likewise deflected. Water rushing past the boat impacts the paddle and swings it to the side, creating amplification of the force and leverage of the vane. This force is transmitted to the boat's wheel or tiller by way of lines and blocks, resulting in movement of the boat's rudder to effect the final course correction.

Pendulum lines (sheets) translate torque from the paddle to the ship's wheel or tiller to effect a course correction. The main priority when installing the pendulum lines is to minimize friction, chafe, and laxity while ensuring convenient inspection of the components. The straightest pathway between servo-pendulum and wheel or tiller is best. Blocks fixed in position are preferred over those temporarily placed.

Excessive slack diminishes the vane's efficiency, but over-tightening prevents blocks from spinning properly and creates friction, noticed especially in light air situations.

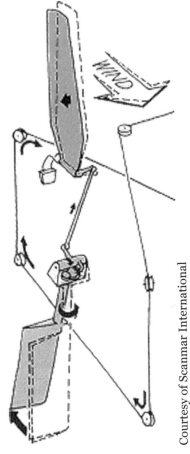

Fig. 6.5
The servo-pendulum unit principle.

Use nothing but high quality roller or ball bearing blocks at least two inches in diameter to minimize friction that hampers performance. Use as few blocks as possible to ensure the most efficient pathway between the mechanism and steering gear.

The lines should be ¼" Spectra, which is extremely strong, resists stretch and chafe, and functions well through blocks in a repeated fashion. Leave some extra line at the forward end during the initial installation to allow for possible eventual chafe or wear after prolonged usage. The line is then slipped a few inches through the leads. Tie a new knot at the pendulum to move defective points of line away from blocks. Mark the pendulum lines at points in reference to specific points on the boat with indelible ink to ensure that you know when the pendulum is centered correctly.

Within the run of the lines, tie a knot similar to a trucker's hitch to create a block and tackle arrangement that allows for convenient re-tensioning of the lines. Make a bowline in the line at the wheel end and a bowline with a 3'–4' tail (the bitter end) in the other end. Leave a gap of about 12"–18" between the two loops. The tail can now go through the first bowline and back to the second and then again back to the first before being locked with a half hitch.

The stainless construction and materials of bearings and bushings make most systems more or less maintenance free. DO NOT put grease into the bearings, as most grease will emulsify or form a hard paste after working together with salt water, causing friction. The bearings and bushings are made from materials that work better with water on them. Maintenance consists of hosing the gear with fresh water to clean out salt deposits or debris.

It is very unlikely that a new wind vane will malfunction while sailing; the sea very seldom causes problems. A quick inspection—*before* you leave the dock—can determine that the equipment is in working condition.

- Check that the air vane moves easily from side to side without friction.
- Loosen the pendulum lines and make sure that the pendulum can swing easily between the legs. The holes on the line attachment on the pendulum should line up with the blocks at the bottom of the legs. If the holes are

forward of the blocks, it indicates that the pendulum has been pushed forward as the boat was most likely backed into a dock or was hit from behind by another boat.
- Check that all the blocks are moving freely.
- Check that the water vane is straight fore and aft when the air vane is straight up. The actuator shaft (connecting the air vane to the water paddle) can be adjusted by twisting either clockwise or counterclockwise, which will make the shaft longer or shorter. If the length of the actuator shaft is not correctly adjusted the *monitor* will not work. This adjustment is extremely important!
- You should consider replacing all plastic bushings and bearings after approximately 15,000 miles or 5 years of use. The old worn parts still work and, as a standard practice, you should save them for an emergency.
- If damaged, stainless steel can be easily welded and worked. Regular hand tools are most often all that is needed to make repairs.
- If possible, you should perform more involved repairs on the gear with the vane dismounted from the hull. It's easy to lose bearings and other parts in the water. There are extensive and very complete manuals accompanying wind vanes, and these should be read and understood not only for installation and repairs, but to maximize performance as well.

Loss of the Rudder

We've covered the steering gear from the top almost to the bottom, and considered self-steering options as well. Now, what to do when the rudder drops off or breaks, and the boat can't be controlled by anything currently in place?

The rudder is a principle element of the Center of Lateral Resistance (CLR), i.e. those components of a boat that prevent her from being pushed sideways by the wind. Other parts of this system include the hull, keel, skeg, and position of ballast within the hull. In combination these elements determine the location within the boat where their individual efforts are concentrated, the CLR.

The CLR resists the combined forces that tend to heel the vessel, known as the Center of Effort (CE). This is composed of the sail plan spars, and windage of the hull. Figure 6.6 below depicts the CE and CLR of a typical sailing vessel. When one component of either center moves or is lost, the whole system changes, and the boat behaves differently.

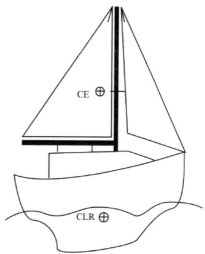

Fig. 6.6
Positions of CE and CLR on a typical sailing vessel.

Loss of the rudder shifts the CLR forward very suddenly. The boat responds with increased weather helm and heads up into the wind. If the boat is on a down wind course when the rudder parts (as it often is because waves from astern lift the transom to free a dislodged or broken rudder), the boat is apt to sail crazily in any direction, driven partially by wave action. She may finally settle into a more or less stable position, depending on the sail plan, sea state, etc., but will swing around at will.

A sailboat can steer herself quite efficiently on a heading anywhere from a beat to slightly before a beam reach in light to moderate wind and wave conditions. I've driven a boat with the sails only for almost 400 miles, using only sail trim adjustments with the rudder amidships. That can also be done without a rudder, although with more difficulty.

The sails must be very balanced with the boat's lateral resistance—that is, the boat tends to maintain course because the CE and CLR are exerting equivalent forces on the hull. First set the jib, appropriately sized for the conditions. Trim it properly from top to bottom, with all telltales flying equally. Next, set up the main sail, adjusting it on the traveler first to position the lower half. Remember that weather helm is increased without a rudder and the CLR forward. Reefing the main and moving traveler position to leeward are common adjustments to decrease weather helm. Next, adjust the sheet so that the main's telltales are flying top to bottom. The upper section of the main sail is fine-tuned by

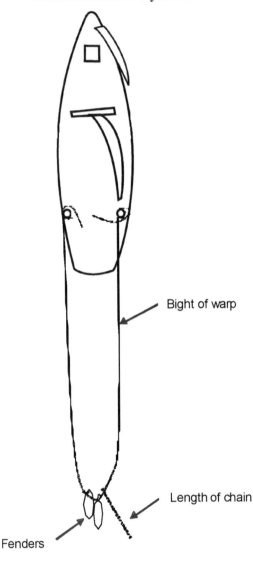

Fig. 6.7
This sample jury rig steering system trails a U-warp aft from a control bridle to stabilize the boat and lend steering control. Note the length of chain to keep the line submerged and pulling.

STEERING

adjusting the sheet, while the bottom sections are adjusted primarily with the traveler car position.

When the sails are trimmed correctly, the boat should sail close to the desired course, heading up and down a few degrees but returning to the average heading. Bring the main up the traveler for more weather helm, down the traveler for less. This technique of sail trim should actually be the normal procedure. It produces a more efficient sail plan, provides less leeway, and relieves tension on the steering gear.

Sail trim alone probably won't be enough unless your course is upwind and your crew is composed of good sail trimmers. One option to cope with rudder loss utilizes equipment that should already be on most boats. A commonly used method is to trail a long U-warp, as depicted in Fig. 6.7 (opposite page), behind the boat to create drag. Steering control is gained by shifting the drag from side to side, directing the boat in that direction. Drag is increased by a number of means: tying multiple knots in the warp, attaching an array of items to the line—fenders are depicted in Fig. 6.7. This, incidentally, is a practical method used to control boat speed when sailing down wind in heavy weather conditions.

The preferred method of gaining steering control is to trail a single warp, (Fig. 6.8 right) rather

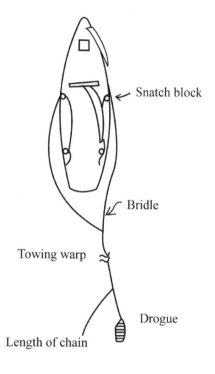

Fig. 6.8
Steering control is enhanced when using this single warp trailing line, which also causes less drag than the bight of warp.

than the bight seen in Fig. 6.7. This enables more precise control of the drag, making steering more efficient.

Each segment of this makeshift system can be comprised of numerous pieces of gear already on board:

Bridle
A suitable bridle can consist of one long line attached in its middle or two separate lines.
 Examples:
- dock lines
- halyards (new or older ones that have been replaced)
- spinnaker sheets and guys

Block along rail
snatch block
single block on lanyard, as in Fig. 6.9

Bridle/Warp connection
single stout ring
large shackle
snatch block with becket

Fig. 6.9
Three feet of Spectra permanently attached to this single block make it portable and useful all around the boat.

Trailing warp
-sea anchor rode
-primary anchor line
-miscellaneous lines tied together

Control weight
- Without some form of weight, most devices will lose efficiency by floating near the surface. Weight, in the form of chain, a small anchor, or an anchor sentinel, keeps the drag device submerged and pulling.

Steering

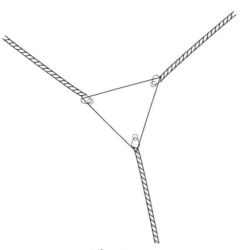

Fig. 6.10
The collision mat is ideally suited to serve as a drag device, since it is constructed of heavy Dacron with reinforced rings at each corner, with strong lines permanently attached. These lines would be brought together, forming a triple bridle to which the warp connects.

-chain
-small anchor

Drag device
-drogue
-fenders
-storm jib
-sail bag
-duffel bags
-tire
-collision mat (Fig. 6.10)

The warp attaches to a bridle rigged at the transom. The bridle is directed forward to turning blocks, from which it leads aft to the cockpit winches for control. The bridle/warp junction must be fabricated to minimize line chafe. Installing a stout ring or shackle at the intersection is ideal (Fig. 6.11, left).

Separate lines are attached to the hardware with bowlines or fisherman's bends, or with a thimble. In Fig. 6.12 (p.

Fig. 6.11
Bridle and warp are seen attached to a large shackle. The warp is trailed from mid-transom to maintain a course straight ahead, as in Fig. 6.12.

Fig. 6.12
The bridle/shackle/warp connection is seen, here positioned amidships to keep a course straight ahead.

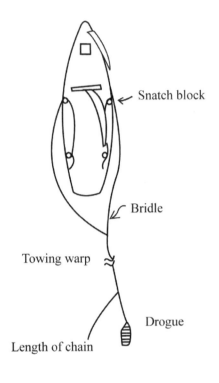

Fig. 6.13
The boat has just turned to starboard in response to a shift in drag to that side.

6.12 (p. 120), the bridle has been tied in its center to a large shackle, and the trailing warp is also in place and ready for deployment. Here, a single loop was used in the warp's knot for short-term deployment. For long term use, all lines attached to a ring or shackle should loop the hardware twice for extra strength. The bridle is connected by taking two loops around the ring and then tying a reef knot on a bight to secure it.

The bridle is maneuvered to either side of the transom to effect a steering change in that direction. Fig. 6.13, above, illustrates a vessel turning to starboard in response to the port bridle being eased as it's taken in on the starboard side. Drag is shifted toward starboard, pulling the stern in that direction so that the boat comes about.

STEERING

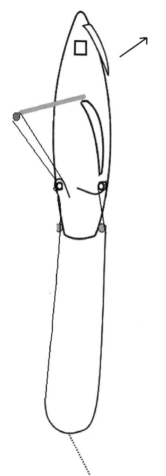

Fig. 6.14
A snatch block attached to this spinnaker pole extends the port side bridle outboard, providing extra leverage and making the system more responsive to steering adjustments from the cockpit.

Steering control with the warp improves even more if we widen the bridle (Fig. 6.14, left) by extending one or both sides outboard, thereby increasing overall leverage of the system. The apparatus becomes more responsive to adjustments of the bridle, with more effective steering corrections. This can be done by leading the bridle through a block on the outboard end of a spinnaker pole, whisker pole, or reaching strut.

Yet another system makes use of gear that comes as standard equipment with the sea anchor: its control line that is normally used to adjust the angle at which the sea anchor rode leaves the bow (Fig. 6.15, following page).

Fig. 6.16, page 123, illustrates the sea anchor's control line, with float and snatch block, attached to this warp that extends from the starboard quarter. The warp is seen here in the amidships position. Steering is controlled by either easing or tensioning the control line to position the warp as desired. Notice in Fig. 6.15 (page 122) that the line has been eased—the warp moves to starboard to effect a starboard turn.

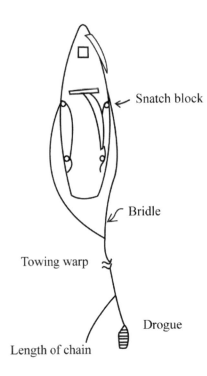

Fig. 6.15
The sea anchor control line's snatch block connects to the trailing warp.

Throwing a replacement rudder together from common boat components has certainly been done; a large oar lashed to the stern or a rudder built from a whisker pole and suitable flat piece of wood are examples.

To Jury Rig a Rudder:

Use any suitable pole-like object on the boat; a whisker or spinnaker pole or reaching strut are examples. Items to use as the blade include a cabinet door and storage lids; these are best attached to the pole with through bolts or self-tapping screws. Holes drilled in the door could also serve as anchoring points for lashings.

If you've made the efforts to prefabricate components for a jury rigged rudder, you're ahead of the curve. This could involve fittings placed on the transom to hold components of a new rudder system.

We have seen that steering control can be lost a number of ways. Problems can arise in any segment of the ship's steering system, including components within the pedestal, the intermediate zone, the rudderpost, and the rudder itself. Self steering mechanisms of all types are subject to wear, deterioration, corrosion, damage, and interruptions in electrical power that lead to more causes of steering failure.

Steering

In many instances, a lack of maintenance factors into gear breaking down. The steering mechanism in general is difficult to access for routine inspection, but responsible sailors understand that this, like all other systems on board, is subject to age, deterioration, damage, and failure. While a loss of steering control is never a welcome turn of events, we should nonetheless have the ability to cope with the tools and materials at hand to replace, repair, or jury rig our way to safe haven.

Fig. 6.16
The sea anchor control line, from a block on the port side transom, connects to the warp trailing from the starboard side.

A true Salty Dog, Dean Gibbons, tends this bridle as we drive Voyager *without the rudder.*

7

Sail Repair

Written by Ed Mapes
with consultation from Travis Blain

This chapter focuses on the repair of sails that incur damage while at sea, away from any sail lofts. Proper care of the sail inventory actually begins far earlier, while the boat is still in harbor. Well maintained sails give better performance and are less subject to wear and damage when in use. Please see the chapter Sails and Sail Repair in *Further Offshore* for a comprehensive discussion of sail maintenance and preparation for use at sea.

Once we've set sail, an important activity each day is monitoring and maintenance of the vessel and all of her systems; the sails are key components and we must be vigilant in our daily examinations, searching for any signs of chafe or tearing that eventually lead to failure. It's important to catch damage early, before it has an opportunity to get worse. This means more than only inspecting in high winds, for sails are at great risk for damage in light air as well. Continual boat motion and lack of constant pressure on the sails actually causes chafe more quickly than heavy weather sailing. Make it a point to check the sails every day that they're in use.

Some of the important areas to concentrate on:

—Leeches for chafing, defective leech cords, batten pockets
—Foot for chafing/tears and foot cord
—Luff tapes
—Luff slides on the mainsail
—Reef cringles and bull rings, grommets
—Stitching of all seams and corners of the sails
—Mainsail headboard
—Any areas of chafe, punctures, or small tears
—Reinforced sail areas at the corners
—Chafe guards

Sails are damaged by a variety of insults. Chafe occurs when the sail rubs against a part of the boat continuously over a period of time. A deck-sweeping jib in contact with the bow pulpit and an eased mainsail impacting a spreader are common examples. Chafe is a type of sail damage that is totally under our control.

Unequal or over-tensioning of sails is the leading cause of tears. This occurs when one section of a sail is tensioned far in excess of other, adjacent areas. For example, extreme outhaul tension, especially with laxity at the leech and luff, can cause long, sail-wrecking tears along the foot of the main. Excessive halyard tension when sailing too high with a spinnaker can contribute to a catastrophic tear of the sail. These are also situations that we should minimize with a thorough knowledge of our sails' construction and understanding proper sail trim.

Ultraviolet radiation weakens sail fabric slowly over time, and can cause failure when sails are placed under heavy working loads. UV damaged fabric takes on a dry, brittle feel, and the material loses its normal luster. Likewise, dirt and salt trapped within the fabric cause damage and degradation on a microscopic level.

Sails should not be folded and stowed when they're wet or dirty. Hose down salt-encrusted sails with fresh water and allow them to dry before folding or furling. Spray soiled sails with fresh water as they're being hoisted, apply mild liquid dishwashing soap, and clean with a large sponge. Once the sails are clean, rinse thoroughly with fresh water and allow to dry before folding.

It is important to discontinue usage of any damaged sail immediately, especially when under heavy pressure. The weakened area is far more likely to worsen, and repairs are easier when the defect is minor. Lower or reef a mainsail so that the damaged area is taken out of service. Lower a jib and haul in a spinnaker when defects are spotted, even small punctures or tears. Before beginning the repair, assess the cause of the damage. Was it from chafe, ultraviolet weakening of the sail material, unequal loading, or being pulled over/across a sharp projectile? It does little good to make repairs only to have the problem recur. Notice the area of damage: high stress or low stress, hardware attachment point, subject to chafe, and so forth. This determines the type of repair, which must aim at preventing future damage. After the repair is done, try to protect the affected part of the sail while it's in use. As an example, after stitching a horizontal tear below the first reef point of the main sail, hoist the sail and tension the first reef line just enough to minimize pressure on the repair area until permanent repairs are made.

Most sail repairs at sea, especially in low tension areas, consist of covering both sides of the affected area with adhesive Dacron tape. Most sailmakers recommend at least 2"-wide sections of tape for best results. Tape doesn't adhere to wet, salty, or dirty material, so begin by rinsing the affected area and then drying. If drying is difficult, apply rubbing alcohol and dry again. Any loose filaments of thread or tattered sail will diminish holding of the tape, and should be removed with scissors or a hot knife. Sail tape may be adequate in some emergency cases, but stitching is far more reliable. Duct tape is even less desirable, and many sail makers will hesitate to repair sails if you've used duct tape. Contact cement may make a strong repair, but the segments of sail must be aligned exactly right because of the instant bonding nature of the compounds.

Sail repairs are best accomplished when the sail is removed from service and taken to a suitable work area with a flat surface and out of the wind (Fig. 7.1 below).

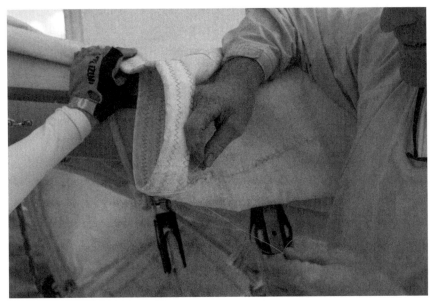

Fig. 7.1
The author attempts to repair a rip in the main sail after lowering the sail on the boom. This proved to be very difficult, so the sail was removed and taken below to complete the fix.

To repair small holes or tears, cut two sections of sail tape large enough to cover the defect and overlap the area by at least two inches on all sides. Oppose the adjacent sections of the sail exactly to avoid wrinkles and an improper patch job if they're not in perfect position. Ask a shipmate to assist by holding the area while the patch is positioned. Place the first section over the defect. Then, while holding it in position, slowly remove the plastic backing from one end of the tape to expose the sticky surface. While the backing is peeled away, press the tape into position. Next, turn the sail over and repeat the procedure on the other side. Once both sections of sail tape are on, press them down and rub firmly for a secure repair. If the defect is in a low-tension segment of the sail, this method can be adequate, but stitching will be needed for larger-sized defects and those in load-bearing areas such as along the foot and leech, as in Fig. 7.2 below.

Fig. 7.2
A tear in Voyager's main sail.

The tear is repaired by placing 2" sail tape over each side of the sail, encompassing the tear after the edges are brought into perfect apposition (Fig. 7.3 following page).

Large holes or tears can be a more challenging fix. If the defect is straight along the sail without jagged edges, sail tape placed on both sides should be strengthened with stitching. The area of tape overlap should also be at least three inches on all sides for more strength. For jagged tears, it's best to replace the torn section of sail with a piece of new fabric.

Fig. 7.3
Sail tape is used to repair the rip in* Voyager's *main sail.

Most tears run parallel to stitching in the material. Bring the sides of the tear into direct opposition, inserting pins if needed to hold the segments in place. Remove the paper backer from one side of double-stick tape, and adhere it around the defect edges on one side of the sail. Align the tape either parallel or perpendicular to the cloth weave direction. Now begin removing the affected area by cutting the sail *inside* the tape area. Use either scissors or a hot knife to obtain sharp sail edges (Fig. 7.4, opposite page).

Lay repair cloth over the sail tape that surrounds the defect, and trace over the outside edges with a pencil. Cut the cloth at the pencil outline and then lay it over the defect after removing the

Sail Repair

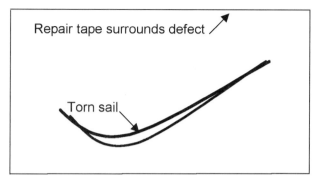

Fig. 7.4
Double-sided sail tape has been placed around the defect, and torn sail has been removed so that patch cloth can be sewn into position.

backing paper from the tape strips. Firmly press the sail down against the strips until all edges are adhered and the repair material is perfectly flat against the sail. The repair section is now sewn into position at the outside edge as described on the following pages, courtesy of *Further Offshore*:

Select white, Dacron/polyester thread for the repair; it lasts longer than dark thread, is UV-resistant, and the end can be terminated by melting instead of being knotted. Always use a triangular-shaped rather than round sewing needle; the hole it makes is less likely to tear. Hand stitch by using a zigzag pattern - ///// - which spreads out the "pull" and stretches as needed. Push the needle through the new fabric and through the original sail. Subsequent stitches are placed at 4–5 mm intervals along the repair. Pull the thread taut at each step, but not so tight as to crimp the sail material. Continue this along the whole side of the repair.

Now stitch backward in the opposite direction, using the needle holes just made, to complete the pattern. Each stitch is pulled taut as before. Terminate the stitching by tying the two ends of thread together, and melt them to form a more permanent closure. Use the same pattern along all sides until the repair is completed. If the repair is in a high-chafe area, cover the stitching with sail repair tape for extra protection.

Stitching along seams stands above the sail, and is therefore more exposed to chafe and deteriorates more quickly than the sail cloth. Eventually, even the best seaming can wear out and allow the seam to part. For repair of worn or torn seams, it is important to oppose the sections of sail perfectly and stitch the sail back together. The existing needle holes simplify alignment, but always use glue--such as Seamstik or a glue-stick from any office supply store--to adhere the two segments together before stitching. Pull the seam apart, removing all original thread material. Next, pull the paper backing from *one side* of double-sided repair tape, and lay it onto the bottom section of sail. Be sure the cloth is perfectly flat on the tape, and press the sail securely into place. Next, press the top sail cloth down onto the sail tape, making sure that it aligns perfectly with the bottom piece, and press it down firmly. This ensures accurate opposition of sail surfaces. Now apply glue to the overlapping seam edges, and press into place.

After the glue has dried and holds the seam edges perfectly, begin sewing the edges together, using the existing needle holes in the sail. Extra-thick areas of cloth may require holes punched in with an awl. Place holes at regular ¼" to ⅜" intervals.

Pressed rings—cringles or bull rings—bear heavy loads, and the sail can tear along one side of the ring. Repair this temporarily by placing 1" tubular webbing through the ring on the side of the defect. This will most likely be along the line of greatest tension. Extend the webbing at least 10" along both sides of sail, using the same stitching pattern as the original reinforcement if possible.

More extensive damage requires ring replacement. Cut out the old ring, preferably using a hot knife for clean edges. Cut double-sided tape, and tape the new ring into position. Extend the tape at least 10" along the sail in the direction of greatest tension, matching the original reinforcement. Now place a second piece of tape in the opposite direction, and further pieces to secure the ring into position, tightly against the sail.

Cut the tubular webbing long enough to stick to a length of tape, go through the ring, and return up the sail to the opposite side. This is done for each piece of tape used. Now sew the sail and both layers of webbing to ensure strength. Use an awl to make the holes, and use a zigzag stitch pattern. Follow this procedure on all sections of webbing to complete the repair.

Many defects on spinnakers are small punctures or tears. No matter how small, immediate repair is mandatory to prevent very serious rips. Anyone who has seen a spinnaker explode understands how a tiny puncture can lead to destruction of the sail. Whenever a small defect is found, douse the sail and make repairs before using the sail again. Your sail repair kit should include a supply of "dots," round sections of sail repair tape. These are perfect to use on these small tears.

Larger tears can also be repaired with sail tape alone, without stitching, on spinnakers. Clean the edges of opposing sections of frayed thread, bring the edges together, and place pre-cut sail tape along one side. Repeat the process on the other side to complete the repair.

These procedures are meant to be temporary, used only to make the sail serviceable until landfall. Any sail that has been damaged and repaired at sea should be taken to a loft for permanent repair.

Recommended sail repair kit contents

Sailor's palm

Seamstick tape, 1/2"

Adhesive Dacron tape: 2" X 2 oz, 3" X 3.8 oz, 6" X 8 oz.

Spinnaker repair tape: 2" X 25'

Waxed nylon thread

Leechline cleats, aluminum

Dacron tape: 3" X 3.9 oz, 6" X 8 oz, 2" X 5 oz,

Replacement slides, slugs, hanks, protectors

Shears, bent blade

Seam rippers

Hot knife

Seamstik glue or glue stick

Sewing needles, assorted sizes, straight and curved

Seizing wire, stainless, 1/16"

Tubular webbing, Strapping, 1"

UV resistant thread: V92

Stainless steel shackles: ¾", 1"

Batten pocket elastic, 1 ½"

3" X 8 oz, 6" X 8 oz.

Leech line Dacron, #505, 1/8", Kevlar, Spectra

Awl, 2 ½"

Wire cutters

Spinnaker "dots"

Preceding material courtesy of *Further Offshore*

8

When the Boat Floods

Taking on water is a scary concept; for many sailors it's one of the most terrifying, with visions of foundering and eventual loss of the vessel (Fig. 8.1). This is the single most common cause for people's fear of offshore sailing, especially when they contemplate sailing great distances from land. Multihulls lack a weighted keel

Fig. 8.1
This vessel lost the battle and slips below the surface.
This is every sailor's nightmare.

and are therefore more resistant to sinking—even if one hull is filled with water. Monohulls are far more at risk because of the heavy keel attached to a single hull; when it fills, the boat goes down.

Water has to enter a boat through an opening in the hull, either one designed to be there or one that we put there by accident. Significant volumes of seawater can ship from a wave flooding the companionway, hatches, ports, or dorade vents, but not enough to sink the boat. Failed through-hull fittings or seacocks, and holes punched into the hull from collisions are the causes of *serious* flooding.

While making one of our passages in the spring of 2007, we sailed in Force 8 winds with 15-20 foot seas from the starboard quarter. I was driving, and had engaged the autopilot to go below. After getting into a conversation with a crewmember, my return to the helm was delayed. While we chatted, a very determined wave picked up *Voyager*'s stern and spun her hard to windward. The next big wave broke over the bows. The onrushing water deluged two dorade vents; from below they looked like fire hoses as seawater spewed into the salon.

After scrambling back up to the cockpit, I noticed that the boat had become inexplicably calm, and I finally figured out that we were very properly hove-to, with *Voyager* riding the oncoming waves in a most controlled manner! The water below was quickly pumped out and the sole wiped dry; we got back on course and continued sailing.

Incredibly, when I was cleaning the boat a few months later, I found the skeleton of a flying fish lodged in one of the dorade vents. The poor guy had been trapped there within the gallons of water that rushed through the opening to the salon.

There are basically three areas of the boat containing pipes and hoses that are important considerations: the fresh water system, the heads, and the engine. Plumbing accidents or failures in any of them don't usually lead to disaster, but can cause considerable repair chores that consume your time and energy.

Leaks in water tanks can lead to loss of valuable water stores. Whenever tanks are filled, look at any accessible areas, and watch for water leaking into the bilge. Access ports can leak small amounts of water, and tank fittings, especially those made of plastic, develop cracks or break off from normal wear and tear. Leaky ports can be serviced by fabricating a makeshift gasket and tightening the lid securely.

Tank fittings can be repaired by drilling new holes through their bases, and then securing with machine screws, washers and nuts, or self-tapping screws. Place bedding compound between the base and tank before tightening.

Whenever a fitting itself breaks, my preferred method of repair at sea is by coating both surfaces with 3M's 5200™, and keeping the segments into the correct position for a day for adequate drying.

Fresh water hoses rarely leak, but it's wise to have replacement hose suitable for splicing if needed.

We know that a punctured hose or one that comes off from a through hull fitting will leak; that's why hoses are always double hose-clamped to provide extra security. Hull fittings themselves can corrode, weaken, and fail though—within just a few years in the salt water environment, much faster if a ground fault electrical defect is at work.

A through hull fitting of 1½ inches in diameter that is two feet below the waterline will admit 70 gallons of water per minute. At that rate, it wouldn't take long for an undetected leak to swamp a sailboat.

Your vessel inspections should always include checking the integrity of the actual fittings by standing on them or rapping lightly with a rubber hammer. Move the seacock handles back and forth to ensure security.

It's also wise to attach properly-sized wooden bungs to each through hull or seacock in the vessel in case of failure. I always kept an assortment readily available in case one was ever needed, though not tied to each fitting. Then an inspector for the Bermuda Ocean Race pointed out to me that were the boat to be capsized, finding the bungs could become impossible, so I've gone back to tying one to each fitting with string through a hole drilled through the tops of the bungs.

Along with regular inspections of all fittings and installing wooden bungs, it's wise to post a hull diagram of the vessel depicting locations of every through hull or seacock in the boat. Every crewmember should be aware of each location. We included fire extinguishers on the diagram used on *Voyager* (Fig. 8.2, opposite page).

Routine checks of the bilge should be commonplace at sea—I recommend a bilge check during each watch. While most bilges have a residual amount of water, you should investigate whenever there's more water than normal in a bilge area. A good first step is to smell the water, trying to distinguish possible causes such as a leaking holding tank, fresh water tank, fuel, oil, seawater, or sewerage. Tasting comes next, which can easily distinguish between fresh and seawater. It is customary for sailboats to be equipped with an electronic, float-switch operated bilge pump in the main bilge area. Larger vessels may have this diaphragm pump in other bilges and in the engine room. These electronic pumps are also activated by closing the breaker(s) on an electric panel. A manual bilge pump, which is capable of expressing far more water than most electronic pumps, is usually positioned on deck, usually in the aft cockpit area.

Installation of a second manual bilge pump below decks is recommended, and is mandatory for ocean racing boats. This pump should *not* be located inside of a cupboard, locker, or any other site that requires opening a door or where gear could cover the pump. A convenient spot is often within the main bilge area, which is usually large enough to accommodate it (Fig. 8.3, page 140).

Voyager

Seacocks, Thru Hulls, & Fire Extinguishers

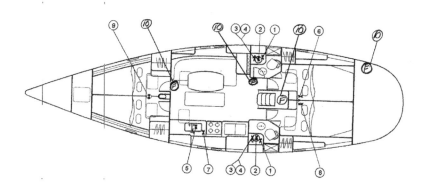

1) Head Discharge
2) Head Intake
3) Washbasin Discharge
4) Shower Discharge
5) Sink Discharge
6) Engine Cooling Water Intake
7) Ice Box Drainage
8) Stern Tube Cooling Intake Valve
9) Speed/Log, Depth Sounder Thru Hulls
10) Fire Extinguishers

Courtesy of Further Offshore

Fig. 8.2
Locations of all through hulls and seacock fittings and fire extinguishers are posted so that all crewmembers know where to look in case of flooding or fire.

A breach in the hull below the waterline is the other cause for concern; boats are most frequently holed by collision with submerged containers and whales that lie on the water surface. At any given time, there are dozens of containers, fallen from the decks of ships, floating just below the surface of our oceans, like the ones in Fig. 8.4 (page 141).

Fig. 8.3
This below-decks manual bilge pump is located in the primary bilge area immediately beneath the removable floor section.

A shipping container is a type of steel storage box used to load heavy equipment and dry goods for shipping purposes. The containers are usually 20 or 40 feet (about 6 or 12 meters) long and 8½ feet (2.6 meters) high, with a pair of doors at one end and a wooden floor. Upon reaching a commercial port, the containers are unloaded from ships and placed on railroad cars or the trailers of 18-wheel semis for further transportation. When a container falls from the deck of a cargo vessel (Fig. 8.5, p. 141, below), it becomes neutrally buoyant, and floats just beneath the sea surface. This makes it virtually impossible to spot in time for evasive maneuvers, and the impact is usually below the waterline. Depending on the vessel's speed at the moment of impact, very serious damage can result, with hull breach and rapid flooding below decks. The same is true of hitting a whale, as in Fig. 8.6, page 142.

FLOODING

Fig. 8.4
This ship has containers loaded in a very typical way. You can imagine these tumbling overboard during storm conditions.

Fig. 8.5
This container ship limped safely into port, but did all of her cargo make it?

Fig. 8.6
We saw this whale within a pod of six; you can understand how difficult it would be to spot one at night. Whale collisions have sunk numerous sailboats.

If a boat slams into a submerged container, whale, or other object, the impact should alert the whole crew to look for flooding at the bow. If the bow happened to be down on a wave at impact, the damage could be above the waterline. If the vessel was rising on a wave at impact, however, the impact would be below the waterline, and the boat is probably shipping water—maybe a lot of water. The crew should:

—Stop the boat, either by heaving-to or turning off the engine.
—Assess the leak by removing anything necessary to gain access to the hole.
—Check the bilge to find out if water is flowing into the boat.
—Begin pumping water with the manual bilge pump(s). The electronic pump should activate automatically when the float switch is lifted by water in the bilge. Inspect the intake hoses frequently; debris can cause obstructions.
—Begin efforts to stem the water flow as described below.

—Note the lat/long position in the logbook and hit the MOB button on the GPS to mark the position there.
   —Communicate the situation to land bases if controlling the flooding is impossible.

If the damage is to the side, rather than the bow, heeling the boat the opposite way can lift the breach out of the water to stem the flow and allow time for repairs.

Holes above the waterline along the bow may be accessed from the anchor well. Damage farther aft will cause leaking behind the anchor well, in the area of the V-berth, double berths, or whatever occupies that space in the vessel. Getting access in this area necessitates removal of bedding, lids covering storage space, sections of flooring, and so forth. This isn't the time to worry about causing damage to those structures; getting access to the leaking area is too vital. Do not hesitate to use saws, hatchets, axes, or hammers to get to the holed section.

Once there, shove items such as deflated fenders, cockpit cushions, life jackets, foul weather gear, or bedding against the hole. Plastic or synthetic waterproof fabrics or material is preferred. If you're able to stop the inrush of water, make provisions to keep that object tightly against the opening. Something flat, such as a cabinet door or section of sole, a companionway slat, or toilet lid, is best. Shore this object tightly in position using objects such as a wooden board, whisker pole, boathook, or emergency tiller. Keep the area open to facilitate frequent inspections; make sure the "plug" stays in place and that water leaking in is controlled as best possible.

After the flooding is controlled from below, a collision mat (Fig. 8.7, page 144) should be positioned along the bow externally over the holed area. Two lanyards are brought aft along the gunwale and secured, while the other line is tied to the stem fitting, keeping the mat directly over the damaged area. Once this is in position, the

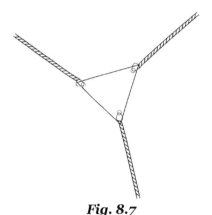

Fig. 8.7
Lines should be connected to the three grommets of the collision mat, ready for use. Two crewpersons pull lines aft along the deck, another holds the forward line. All lines are secured to deck cleats to keep the mat in place over the holed area.

boat can continue to the nearest port for repairs. When the vessel gains way, water pressure against the hull actually presses the mat more firmly against the hole to help create a better seal. Positioning the collision mat is easier and done most accurately when a swimmer guides the operation from the water. This is only done when wind and seas permit, and the person wears a life jacket and is tethered to the boat. If at night, deck crew monitors the swimmer with flashlights.

Leakage from a through hull or seacock can be more dangerous than collision damage; water can flow into the boat unbeknownst to the crew until floorboards start lifting off the sole. By this time, there can already be a significant amount of water in the boat, and locating the leak becomes more difficult. Consult the hull diagram if necessary to locate all openings in the hull, but begin inspecting every one immediately. The steps to take:

—Stop the boat by heaving-to or shutting down the engine.
—Immediately begin pumping water out.
—Immediately begin to inspect for a broken seacock/through hull or displaced hose. Do not forget the engine room, packing gland, and instrument transducers.
—Use a wooden bung to plug the hole when located.

It is possible to use the engine's raw water intake hose to rapidly pump water from the boat. However, given the time it takes to set this up, it is usually more prudent to begin hand pumping first. If there are enough crewmembers available for one to set up the engine pump, that would pump more quickly. To use the engine as a flood water pump:

—turn off the engine.
—shut the intake seacock.
—remove the hose from the intake seacock.
—connect that hose with another that extends to a bilge area containing flooded water.
—run the engine to pump the water.

A very prudent step is to install a three-way valve in the engine's raw water intake hose, between the intake seacock and water filter. Converting to bilge water would then be only a matter of shutting the seacock and turning the valve toward the bilge hose.

Remember, the flood water is now being used to cool the engine. The intake hose must be constantly monitored to prevent obstructions. If the intake hose is obstructed, the engine will overheat, further complicating the situation.

Whenever a boat takes on water, it's considered a serious situation; quick, decisive crew action is necessary, either to minimize damage to contents or prevent the boat from sinking. With a little forethought you can prepare a hull diagram, make the crew aware of through hulls, etc., place wood bungs where they're accessible, familiarize yourself with possible objects useful in stemming water flow, and think about how to deploy the collision mat. These very helpful precautions require little effort; also, the prudent mariner will discuss such topics with the crew before sailing, especially on a long passage.

All efforts must be exhausted in controlling a flood situation. The skipper should be decisive and fully versed in the procedures necessary. Several important tasks will be going on simultaneously, and delegation of those tasks to the appropriate people is important. As in any emergency, it is very important to keep a cool head and avoid panic; crewmembers feed off of and copy the demeanor of the person in charge, and maintaining one's wits in a crisis is very important.

If locating or stemming the inrush of water is impossible and water flows faster than pumps can remove it, a decision must be made to prepare to abandon ship (Fig. 8.8, below).

The crew should deploy the life raft, and be ready to abandon ship, but remain on the boat until all hope of salvage is exhausted. Remember, it's better to stay on a swamped vessel than to crowd into a life raft. Please see the Abandon Ship chapter to learn the correct techniques for leaving the boat.

Fig. 8.8
The crew of this trimaran remains with the vessel, which is in no danger of sinking.

9

Fire at Sea

There can be no more devastating danger to a vessel afloat than an uncontrolled fire. Though fires are surprisingly rare on sailing vessels under way, their detection calls for immediate response to prevent the fire's spread, damage to boat systems, and possible catastrophe.

Marine insurance statistics, seen below, illustrate and rank the causes of boat fires and their frequency. This information is useful in making us aware of the preventative measures, equipment, and tactics useful in fighting these fires.

1) AC and DC wiring/appliance 55%
2) Engine/transmission overheating 24%
3) Fuel leaks 8%
4) Miscellaneous 7%
5) Unknown 5%
6) Stove 1%

These figures demonstrate that electrical wiring and appliances are by far the leading cause of boat fires, and deserve our attention to prevent and extinguish fires of this class. Overheated DC engine voltage regulators are responsible for the lion's share of electrical fires, even though they are thermo and overload protected. Electrical fires are often the result of short circuits, especially at exposed battery terminals, but also within wiring or switches. These are often related to chafe (Fig. 9.1 below) and corrosion of electrical components. The inappropriate use of undersized wires, often installed by amateur electricians, is another common cause of boat fires. Other sources include bilge pump wires (also indicted as a leading cause of ground faults), and instrument wires chafing on hard objects such as vibrating engines or sharp-edged bulkheads.

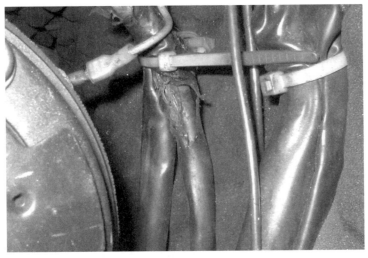

Fig. 9.1
Although it had been conscientiously wire-tied to prevent motion, this tie actually caused the bare wire seen above. This is an example of chafed wires that lead to short-circuit fires.

Shore power AC is an area of increased risk as well; 11% of fires were started by AC systems, commonly at the shore power inlet. AC heaters and other household appliances brought onto boats are also involved in marine fires each year. Electrical fires can be hard to put out because the source of the heat (a shorted wire) can reignite even after a fire extinguisher has been used. Boats must have a main battery switch and/or AC breaker to turn off the boat's entire electrical system.

Overheating of engines and transmissions causes approximately one-fourth (the second highest rate) of boat fires, a statistic that was surprising to me because overheating is signified by an alarm that alerts us to shut off the engine to prevent further damage. However, impairment of the raw water cooling system leads to heat buildup in the engine, which melts hoses, impeller, and even the muffler.

Fig. 9.2
Gasoline fires burn hot and fast, producing thick, acrid smoke that makes them difficult to battle.

A hallmark of engine room fires is black, acrid smoke from smoldering/burning hoses, making the source easy to identify. These blazes are often just smoldering rubber until someone opens the engine compartment and allows fresh air to enter, igniting the actual fire. Engine room fires are most effectively doused by an automatic extinguisher system in the engine room. The next best avenue, after the engine is shut off, is to use an extinguisher in a fire port, or to crack the hatch as little as possible to use the extinguisher.

Any sailing vessel still using a gasoline engine, or storing gasoline aboard for the dinghy engine, runs the risk of the worst kind of fire to have on a boat; 95 percent of fuel-related fires are caused by gasoline, as in Fig. 9.2, previous page. These fires burn fast and hot, and are characteristically difficult to put out. Typical problem areas are fuel lines, connections on the engine itself, and leaking fuel tanks.

Fortunately, the first warning sign is usually the smell of gas that is easily detected—if you can smell raw gas, something's wrong. The majority of gasoline fires actually occur at the dock, during or just after refueling. Steps to prevent fires or an explosion when refueling with gasoline include the following:

- Refill portable gasoline cans on land, away from the boat.
- Shut off the outboard engine.
- All passengers should leave the vessel.
- Have paper towels at hand to clean up all spilled fuel.
- Extinguish all flames, including cigarettes.
- Shut off battery switches; do not use electrical appliances.
- Carefully fill the tank, avoiding spillage; do not overfill.
- Screw on the tank lid very securely.
- Do not start the engine until all traces of gasoline odor have diminished.
- While sailing with the dinghy on board, stow the gasoline can on deck or in the secured dinghy to prevent spilling and escaping fumes.

Fires involving the cooking stove are much less frequently seen than in the past, due in part to the reduced use of alcohol stoves, but also because of strict federal regulations governing the storage and delivery of fuels to galley stoves, heaters, and the like. Rules regulating gas lockers, hoses, regulators, and solenoid valves have made great improvements in galley safety.

Flammable liquids still pose a serious fire hazard, though; they'll not only ignite but can also explode. Propane and butane (LPGs,) and compressed natural gas (CNG) are the most commonly used cooking fuels on sailboats. Both types are odorless, so substances are added to give them the smell of sulfur. LPGs are heavier than air, and sink to the bilge where they accumulate and pose the risk of explosion if ignited by a spark or flame. Leaked CNG rises to the headliner, but can still burn and explode. A propane "sniffer" should be installed with the sensor either low (LPGs) or higher (CNG) in the galley to detect any leaked fuel and sound its alarm when present.

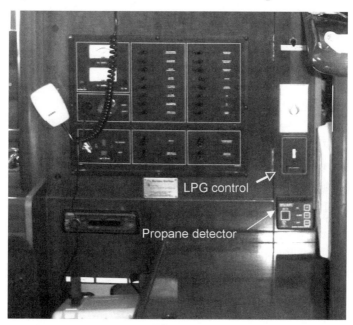

Fig. 9.3
Voyager's LPG solenoid control switch (upper arrow) and propane sniffer (lower arrow) are pictured at the nav station.

Correct usage of LPG stoves also helps to prevent fires. Switch on the propane sniffer to detect the presence of any leaked fuel before engaging the solenoid to start the flow from the tank. LPG tanks must be positioned in a designated locker on deck and isolated from the boat's inner compartments. The tank must be fitted with a shut-off solenoid, controlled from below as in Fig. 9.3 (preceding page), to manage the flow of fuel. Another manually-operated valve should be installed in the galley near the stove. Both valves must be turned on before gas can reach the stove.

It's also important to shut the flames off correctly after cooking to bleed all gas out of the line, from tank to stove. When finished with the flame, the first step is to turn off the external solenoid; the flame will continue burning until all gas in the line is burned off. Next, shut off the manual valve, and then the stove control knob(s).

Diesel fuel is less flammable than gasoline, but it does burn. Most significantly, when leaked from a fuel line or tank, it follows the hull's contours toward the bilges and can ignite in a location other than where it originated. This is also the cause of possible re-ignition, since it can be difficult to locate all of the leaked diesel.

Whenever these liquids are ignited, they burn very hot, produce noxious smoke and fumes, and can set fire to other flammables on board.

The threat of a shipboard fire, especially at sea, is a contingency that all crews should understand. Everyone on board should be made aware of fire extinguisher locations and proper techniques for fighting fire.

Knocking the fire down rapidly is especially important, since evacuation from that confined, flaming, smoke-filled space means entering the water. The protocols we've established on *Voyager* are included in our predeparture orientation program for all crewmembers:

Voyager Fire Protocols
- The person detecting the emergency yells "**FIRE**" to alert the crew.
- The boat is maneuvered so the flames are blown to leeward.
- All hands should know the locations of fire extinguishers from our discussions and the boat schematic.

- Fire extinguishers are manned to fight the fire at once.
- If the fire is in the engine compartment, the engine is turned off immediately.
- Propane tank is turned off at the tank.
- Electricity is turned off at the battery isolation switch or battery posts.
- Flaming materials are thrown overboard if necessary.
- If fire is uncontrollable, begin Abandon Ship protocols.

The fundamental principle of firefighting is to negate one of the essential components for burning to occur: a flammable substance, a source of ignition, and oxygen. Fires need all three components; remove one and you'll defeat the flames. Fire extinguishers are very effective, using a variety of methods to remove oxygen from the flames.

Fires are rated as Category A, B, C, or D; extinguishers are designed to fight fires according to those designations. We're only concerned with the first three categories; D extinguishers are used in laboratories. Each fire extinguisher also has a numerical rating that serves as a guide for the volume of the container's contents; the higher the number, the longer it will dispense its contents.

Class A extinguishers are for ordinary combustible materials such as paper, wood, cardboard, and most plastics. I remember this class because the fires result in Ashes. The numerical ratings on these types of extinguishers indicate the amount of water they hold and the amount of fire they can extinguish.

Class B fires involve flammable or combustible liquids such as gasoline, kerosene, grease, and oil. The numerical ratings for class B extinguishers indicate the approximate number of square feet of fire they can extinguish.

Class C fires involve electrical equipment, including appliances, wiring, circuit breakers, and outlets. *Never use water to extinguish class C fires--the risk of electrical shock is far too great!* Class C extinguishers do not have a numerical rating. The C classification means the extinguishing agent is non-conductive.

Class D fires involve flammable metals that are used primarily within the laboratory setting. These are of no concern for mariners.

Many fires will involve a combination of these classifications. Most fire extinguishers on boats have ABC ratings, indicating what class of fires they're intended to extinguish. Here are the most common types of fire extinguishers:

• Water extinguishers or APW extinguishers (air-pressurized water) are suitable for class A fires only. Never use a water extinguisher on grease or electrical fires, for the flames will spread and make the fire worse. *Water extinguishers should not be on sailing vessels because of this narrow spectrum of utility and danger in use against type B and C fires.*

• Dry chemical extinguishers come in a variety of types and are suitable for a combination of class A, B, and C fires. They are filled with foam or powder and pressurized with nitrogen. A key feature of dry chemical extinguishers is that the residue helps to prevent re-ignition of fires. This residue, however, is corrosive to metallic boat components, and should be cleaned as soon as practical after the fire.

FIRE AT SEA

- BC This dry chemical extinguisher is effective to combat flammable liquid and electrical fires. It is filled with sodium bicarbonate or potassium bicarbonate. The BC variety leaves a mildly corrosive residue that must be cleaned immediately to prevent any damage to materials.

- ABC This is the multipurpose dry chemical extinguisher (Fig. 9.4). The ABC type is filled with monoammonium phosphate, a yellow powder that leaves a sticky residue that may be damaging to electrical appliances such as a computer. Whenever a dry chemical extinguisher is used, the fire will probably be contained but you can count on a real mess to clean up. It's important to start wiping up the chemical right away, because the damage it causes increases with time.

Fig. 9.4
The A,B,C type of dry chemical fire extinguisher most commonly seen on boats.

- Carbon Dioxide (CO_2) extinguishers, used for class B and C fires. CO_2 extinguishers contain carbon dioxide, a nonflammable gas, and are highly pressurized. They don't work very well on class A fires because they may not be able to displace enough oxygen to put the fire out, which allows re-ignition. Exercise caution not to use CO_2 extinguishers in enclosed areas, since they displace

155

oxygen we need for breathing. These are well suited for use in a closed engine room fire. CO2 extinguishers have an advantage over dry chemical extinguishers since they don't leave a harmful residue.

Adequate distribution of extinguishers about the boat is an important consideration; they should be spaced out in all sections of the vessel. There should be access to an extinguisher no matter where a fire erupts. Units should be in plain view and easily accessible at a moment's notice. On *Voyager*, we have one forward at the starboard berth entrance, another is to the left of the galley stove, and another is located in the navigation station. I secured one more in a cockpit locker on deck. The locations are all identified on a hull schematic posted in the salon for crewmembers to see. That schematic also locates all through hulls and seacocks in the boat.

The National Fire Protection Association (NFPA) recommends:

Inspection of the ship's extinguishers at least monthly to ensure proper function. If the extinguisher is damaged or needs recharging, get it replaced immediately!

The pressure be at the recommended level. On extinguishers equipped with a gauge (such as that shown on the right) that means the needle should be in the green zone--not too high and not too low.

The nozzle or other parts not be obstructed.

The pin and tamper seal (if it has one) be intact.

There be no dents, leaks, rust, chemical deposits, or other signs of abuse/wear. Wipe off any corrosive chemicals, oil, gunk, etc., that may have landed on the extinguisher.

Fire extinguishers should be pressure tested (a process called hydrostatic testing) after a number of years to ensure that the cylinder is safe to use. Consult your owner's manual, extinguisher label, or the manufacturer to see when yours may need such testing.

If the extinguisher is damaged or needs recharging, get it replaced **immediately**!

Courtesy National Fire Protection Association

Operation of different fire extinguishers is fairly standardized. Once the extinguisher is released from its bracket, pull the pin at the top of the nozzle mechanism. While aiming at the **base** of the fire, squeeze the trigger to release the powder. Carefully and slowly spray from side to side until the flames are out. Most extinguishers spray their contents for only about 10 seconds, so work efficiently. After the fire is extinguished, back away while maintaining eye contact with the area.

Once the fire alarm is sounded by the alerting crewmember, all hands are called upon to help control the blaze. The most immediate measure taken is to man fire extinguishers to control the flames. In addition, the boat's cooking fuel canisters should be turned off at the tank, the engine stopped after carrying out any necessary maneuvering, and batteries shut off at the battery isolation switch or primary bus cables. Whenever Type A materials such as bedding, seat cushions, or sails are involved in the flames, the best course of action may include throwing those burning materials overboard to rid the boat of that flame. Remember, most of the components of vessels are flammable, and rapidly removing as much flame as possible can be crucial.

If the fire is on deck, maneuver the vessel so that the flames are directed to leeward, and remove other flammables from the burn vicinity.

It's important to identify and eliminate any hot embers or ashes remaining after the primary fire is extinguished and to eliminate them from the boat. Further dousing with water is also recommended.

In a serious fire that threatens the boat, establish communications by whatever means is indicated, relating the vessel position and characteristics, nature of the fire, number of personnel on board, and specific threat to the vessel. If the fire proves to be uncontrollable due to flames and/or smoke, if an explosion is involved, or if sinking becomes imminent, discontinue efforts to save the boat and concentrate on abandon ship procedures to save the crew.

10

Crew Overboard

Along with "We're out of food!" there's no statement that would send chills down my spine quicker than "Man overboard!" We spend a lot of time teaching students how to retrieve a person in the water, but the premier rule that we lay down on *Voyager* is: **Nobody goes overboard; *ever*.**

Prevention is the name of this game, and we strive to make it impossible for an overboard situation to ever happen. My attention was keenly focused on this tenet the first time I sailed *Voyager* solo, from Tortola to Annapolis, in 2003. Solo sailors are most assuredly goners if they fall overboard, and I concentrated on developing protocols to prevent it. Now I apply those same rules to everyone who sails with us:

1) Nobody leaves the cockpit without being harnessed in.
2) Nobody goes to the foredeck alone at night.
3) Harnesses are always worn in limited visibility (night, rain, fog).
4) The Overboard Alarm is worn by the watch at night and during rough weather.
5) Boys (and girls for that matter) urinate below, while *seated* on the head.
6) Don't break rules 1-5.

If we make it impossible for a person to go over the side, we'll never have to face that critical rescue situation. Jacklines are essential elements in this strategy. They provide a safe attachment for tethers as they extend from about four feet from the bow to just before the cockpit area. If the boat is a center cockpit design, jacklines may be required on the aft deck as well, depending on how much room there is. Fig. 10-1 demonstrates that *Voyager*'s jacklines, from their secure attachment forward, run on either side of the boat, inside of the shrouds on either side of the mast, and inside the mast pulpit.

Fig. 10.1
***The deck jacklines are still attached firmly to* Voyager**
as she lines quietly in St. George's Harbor, Bermuda.

Keeping jacklines inboard at all times prevents a person's tether from extending past the lifelines, and avoids the possibility of their becoming chum while being dragged from the boat while dangling in the water.

On ocean races, vessels are required to install cockpit jacklines as well; they enable the crew to attach a tether immediately upon entering the cockpit from below, and to be secure whenever they're in the cockpit in rough conditions. These, just like deck jacklines, should be made of strong webbing attached to securely mounted pad eyes with adequate backing plates (Fig. 10.2). This is a great idea for cruising vessels as well, and would be a great addition to the safety gear on any boat.

Use strong webbing for the jacklines—it's flat and won't roll underfoot. Wet the material before running the lines; the webbing becomes tighter when it dries and won't loosen as much when decks are awash.

Fig. 10.2
One of Voyager's cockpit jacklines. The webbing is attached to very secure pad eye, and droops to floor level so that nobody can trip over the lines.

Another line of defense, and one of the best gear additions I've ever made to *Voyager*, is the overboard alarm. Shipmates carry one of the transmitters at all times during watches, at night and in reduced visibility, and during heavy weather. Even if a person on deck did break several rules and go over the side, the rest of the crew would know it instantly when the alarm sounds.

In that event, well prepared crews know what to do because the protocols have already been explained to them. This discussion has to be part of pre-sail training. Reactions will be quicker, and the sense of panic that can sweep across the boat like a rogue wave can be avoided.

Here are the protocols we use for victim retrieval during our offshore passages:

Crew Overboard Procedure

- The first responder should yell **"CREW OVERBOARD"** to call all hands on deck.
- One person is designated as the "spotter" who POINTS TO THE VICTIM, never taking eyes off them.
- Deploy the MOB POLE and/or Automatic Floating Strobe Light as close to the victim as possible.
- Enter MOB on the GPS, and write down the lat/long position. Calculate the boat's retrieval (reciprocal) course to the victim.
- Place the boat on autopilot STANDBY or disengage the wind vane, and have the ablest person take control of the helm and direct sail maneuvers.
- The QUICK STOP method is initiated as the first preference. The second option is the DEEP BEAM REACH tactic.
- Heaving-to immediately is also a viable option, especially when sailing downwind and when flying a spinnaker.
- Deploy the LifeSling® as the boat maneuvers.
- Approach the victim to leeward, maneuvering the LifeSling® close.
- Luff sails, check the transom for lines, and start the engine with transmission in NEUTRAL.
- Retrieve victim by any means necessary.

—Courtesy of ***Further Offshore***

The person witnessing a shipmate going over the side becomes a crucial element in their rescue, performing two vital functions:

1) Raising the alarm as a call to all hands on deck.
2) Maintaining eye contact with the victim, never taking eyes off of them until another crewmember has them in sight and assumes that job. From then on, that person's only job is to watch and point toward the victim until the boat draws near for retrieval.

A life ring with automatic floating light and/or MOB pole should be heaved into the water as near to the victim as possible. This is especially critical during reduced visibility conditions; a person in the water is a surprisingly elusive target in the best of conditions, but when obscured from view by waves, darkness, heavy rain, or fog, they become even harder to spot.

Most lanyards on throwable devices are only 50 to 75 feet long; a boat traveling at 6 knots covers 10 feet per second. Do not tie the bitter end of throwable devices to the boat because the person will be out of range within 10 seconds, even if you're only sailing at 3–5 kph. Throw the device as far and as close to the victim as possible, and understand this: it won't travel very far if you're throwing it into the wind.

Someone nearest the GPS must immediately press the **MOB** button so that it records the lat/long position, and then *write down* that position and course. Reciprocating that heading (heading ± 180) should take the boat at least to the vicinity of the victim, but wave action and current immediately begin to alter their location. Take note of the wind direction and strength, along with current set and drift to help in estimating their changing position. This information gives an immediate return course.

The person most adept at steering the boat and sail handling should assume helm control and direct rescue maneuvers. The return procedure depends on what has been practiced the most and the states of wind and sea.

If the person just fell off the boat, their drift shouldn't be a big factor. The safest maneuver, when the heading was anywhere from a beat to a reach, is to immediately tack. Tacking is the first step in the quick stop maneuver (Fig. 10.3) or heaving-to, and is safer for crewmembers than gibing on short notice; you don't need more confusion from an out-of-control boom. Reacting quickly is also an enormous boost to the victim who sees the boat maneuver quickly, rather than continuing to sail in the opposite direction. I can attest to this personally, since I've been that person waving his arms while treading water as the boat continues to sail farther away. Believe me when I tell you it's a *sinking feeling!*

Make the tack, and if at all possible and the seas are moderate enough to allow, execute the Quick Stop Technique.

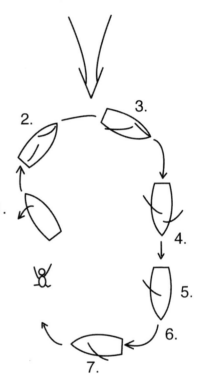

Fig. 10.3
The Quick Stop Technique
***Courtesy of* Further Offshore**

1. Tack
2. Begin turn downwind; ignore jib if crew not available to tend.
3. Continue turning to directly down wind. Ease out main sheet.
4. Sail down wind, trimming sails if possible, but not necessarily. Drop jib if possible.

5. Control the main sheet to prepare for jibing.
6. Jibe the boat.
7. Sail upwind to leeward of victim. Luff or drop all sails.

This represents the fastest means for returning to a victim, always keeping the boat relatively close by. After the tack, the boat heads downwind on a near reciprocal course. The jib need not be tended at this time, but the main should be trimmed. While sailing downwind, prepare the main sail for jibing. After sailing down for only a few boat lengths, gibe the boat to complete the circle, taking it to a close reach toward the victim. The jib can be luffed or rolled in at this point. When the boat reaches a position downwind of the victim, it's luffed up high, and should be able to coast to a position just to leeward. All sails should be luffed or dropped at this point as the final approach is begun.

I recommend approaching the victim to leeward for two reasons: Anyone with experience driving a racing yacht knows that as the boat approaches a mark to leeward, the helmsman loses vision of the mark because the jib is in the line of sight, and a spotter is needed to call out the approach. The same holds true when approaching a victim in the water. Secondly, boats are affected by wave action and driven to leeward at a faster rate than a person in the water; a vessel to windward is apt to be driven down onto the victim. Remember also that the bow moves to leeward a lot more than the stern, so an approach with the bow to windward leaves the helmsperson blind to the victim's position with a bow that tends to be driven downwind by the waves. Not a good situation if you're the guy in the water!

The Deep Beam Reach (Fig. 10.4, page 166) is also useful when the course was upwind to a broad reach. It involves sailing a broad reach to a position downwind of the victim, then tacking to make the approach upwind.

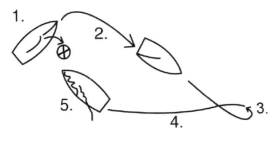

Fig. 10.4
Deep Beam Reach
***Courtesy of* Further Offshore**

1) Fall off to a broad reach.
2) Trimming the main sail for speed, furl or drop jib.
3) Tack.
Sail a beam reach toward the victim.
4) Make final approach to leeward.

The maneuver begins by steering downwind to a broad reach, trimming the main sail for speed as the course leads to a point two boat lengths past the victim. The crew can furl the jib or drop it. Next, tack the boat, and sail upwind toward the victim. Luff the sails and roll in the jib when reaching a position close enough to head up and coast toward the victim, coming to rest to leeward if possible.

These two techniques are recommended except when the boat is headed downwind, especially when traveling fast with a spinnaker flying. Tacking is not an option in that situation, nor is continuing to barrel along while the distance from the victim increases rapidly. We teach crewmembers to immediately heave-to when sailing down wind, essentially stopping the boat and giving crewmembers time to throw things to the victim, shout assurances, and get the boat settled. The maneuver does in fact tack the boat, which is headed up and through the wind to the other board. The mainsail, though, is controlled on the traveler and the helm steered up, allowing the boat to maintain itself on a heading that approximates a close reach. Wind will blow the spinnaker back toward the vessel; ease the sheet, guy, and halyard and take it down as fast as possible. Once the spinnaker is handed, ensure that no sheets/guys are in the water, start the engine, and use it to maneuver the boat to the victim. Make your turn, circle back around the victim, and make the approach from leeward.

Whether the vessel approaches the victim—be that with the sails, as in the quick stop maneuver, or with the engine—it should be possible to release the LifeSling and maneuver its lanyard to within reach so he/she can grab it (Fig. 10.5, below).

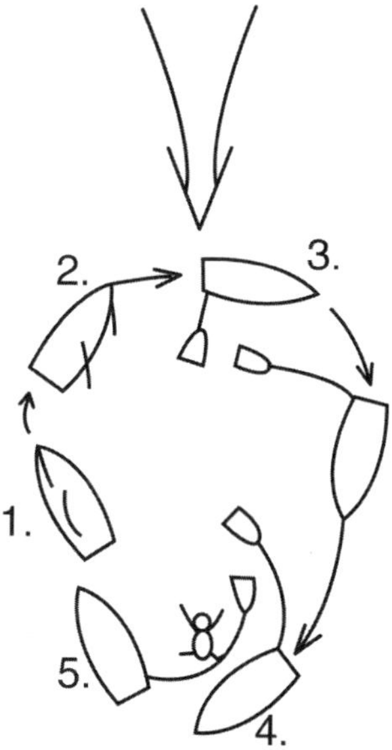

Fig. 10.5
This boat is maneuvered with the engine to bring a LifeSling close to the victim. The device should be mounted permanently at the transom, with the bitter end of the lanyard tied to the stern pushpit. When needed, just open the Velcro tab on the lid, remove the sling from the bag, and toss it into the water aft. Courtesy of Further Offshore.

Bear in mind that the transmission must be in neutral when you've drawn near to the victim whenever the engine is running. Also, monitor the position of the lanyard so it doesn't foul the propeller.

Don't hesitate to encourage use of the engine during rescue maneuvers, especially during limited visibility and when wave action makes sail maneuvers difficult or impossible. It takes practice to master the techniques under sail, and doing them well when a person is in the water is even tougher. Teaching maneuvers under sail in moderate winds and waves of an inland lake or bay is only one side of the coin; executing those maneuvers in rough conditions at sea is another entirely.

I would encourage skippers to practice the overboard rescue maneuvers with their boat not only under varying conditions, but also at night. Making correct turns without the visual input of wind/wave direction and surroundings is far more difficult than one might imagine. Those familiar with night sailing have an advantage, but even then, they should practice these techniques after dark.

A spotlight should be in a convenient location in the cockpit whenever sailing in restricted visibility. I have mounted a 12-volt DC outlet on the cockpit pedestal for ready use of the spotlight. The spotter should shine the light directly at the victim during limited visibility, and never take it off until the ship is alongside.

If a person falls overboard without being seen or heard, there will be a greater—maybe much greater—distance between boat and victim. When the situation is discovered, immediately hit MOB on the GPS and write down the current position and course. If a plotter is available, plot a course to retrace that on the screen. Determine the current heading, wind direction/speed, current set/drift, and wave/swell directions. The person is assuredly downwind, down wave, and down current from your position.

Calculate your reciprocal course or retrace the plotter course line, and prepare to make that course. Execute a sharp turn to leeward—a windward turn takes you in the wrong direction—and begin sailing/motoring your calculated course. Figure out the most likely time frame that the victim could have fallen overboard, and how long ago that was. Now multiply that time (in hours) by the speed (in knots per hour) you've made during that time. This gives the approximate distance (in nautical miles) that you've covered since the victim fell overboard. Estimate the distance the person could have drifted in the same elapsed time.

The crew, using all the binoculars on board, takes watch positions around the boat, scanning the horizon for signs of the victim. Sail/motor the reciprocal course until you've covered the maximum

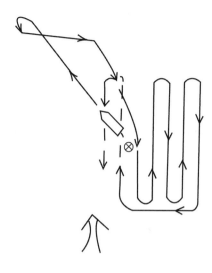

calculated distance you could have traveled from the victim. At that point, make another turn to leeward, and reciprocate the course again. The search should be concentrated on the most likely area for victim location. Use the search pattern as shown in Fig. 10.6. Make successive turns down, trying to maintain a more or less windward/leeward configuration to reduce roll motion.

Fig. 10.6
The Rescue Search Pattern takes the vessel from windward to leeward to decrease rolling motion. Courtesy of **Further Offshore**.

When the boat reaches a victim in the water, the most difficult part may still be ahead—retrieving them back on deck. This can be relatively easy in moderate sea states with a conscious and uninjured victim who can actively take part in the retrieval. In that situation, if the boat has a swim platform and ladder, you may only need to maneuver the transom close enough for him/her to climb back aboard. That is often the case if the person was seen falling overboard, and the boat responded quickly.

The odds of a healthy victim diminish rapidly when the water is cold or rough, if they're injured or panic, or if they're not wearing a PFD. An incapacitated person unable to assist can be exceedingly difficult to rescue. The rescue will almost certainly have to entail:
1) A crewmember in the water.
2) A method to use mechanical advantage to hoist the person onto the boat.

Putting another person in the water is done only when absolutely essential. First, get a LifeSling or other apparatus to the victim. If they can put that on despite their injuries, cold, or fatigue, you may be able to retrieve them without going into the water. If they're unconscious or can't assist for any reason, it will be necessary for the ablest person on board to give direct assistance.

Before entering the water, shed extra bulky clothing and boots. The clothing does not keep cold water out or retain heat, but it does decrease buoyancy by absorbing large amounts of water and makes maneuvering more difficult. Put on a PFD, and connect a tether long enough to reach the victim. Other crewmembers are stationed on deck, attending the tether, watching and ready to help in any way.

The sun shone brilliantly on Lake St. Clair, in Michigan. The fall race series there is always chilly, and the wind, though cold, is great for racing. This was the final downwind leg; our ST7.9 was leading our class, going fast and headed for the line. When the lifeline snapped, Jeff, our foredeck guy, instantly fell into the 51-degree water because he was really hiked out. The helmsman brought the boat head to wind, and we stopped, but about 60 feet from Jeff. He was thrashing wildly, and crying out for us to help. I began throwing cockpit cushions toward him, but the wind caught each one and blew them right back toward us. While Jim and his friend Dave struggled to bring the kite down, the waves took us farther away from Jeff, who was now pleading for us to hurry. I decided to do one of the dumbest things ever: with all of my warm clothing on, no life jacket, and no tether, I jumped in and swam for a fender I'd already tossed in. The intent was to swim it over to Jeff until the boat made it back.

When the cold water hit, it sucked the air right out me, and I was instantly gasping for air as the Mammalian Diving Reflex took control. My soaked, heavy clothes (Patagonia top) began pulling me under, and I had to kick hard to keep myself above the waves. Swimming toward the fender was almost impossible, as I was fighting just to keep my head above water. I finally did get there, and began making slow progress into the waves toward my friend. We were both in trouble now; our boat was still not sorted out, and another boat sailed right past us! I was still 25 feet from Jeff, who had stopped calling out, and was barely able to keep his head up; exhaustion and the cold were taking a heavy toll on him.

Another racer, from a bigger class that had already finished, had seen the trouble and was heading for us. Case, a veteran sailor, removed his clothing (down to his underwear!—not a pretty sight, I can assure you) and was preparing to dive in. The boat drew near, and two hands reached toward Jeff and hauled him up over the lifelines. They pulled me up the side and I grabbed the toe rail and swung a leg up. I was on deck in another few seconds.

I made a lot of mistakes that day, and only made the rescue situation worse by not helping Jeff and becoming a victim myself. In these situations, it's imperative to forego instincts of throwing caution to the wind; stick with the rules that you know.

Lifting a limp person up over the lifelines and onto the deck is usually more than one and sometimes even two people can do by hand. Freeboard on many boats is more than an arm's length, so even reaching a victim from the deck is impossible. It usually takes a line attached to the person or to an apparatus around them. That line is most easily hoisted using the power of the boat's winches. A device such as the LifeSling, a harness, or a ring buoy can also be fitted onto the victim, and the hoisting line attached to that. The line itself can be attached to the victim using a knot such as the French bowline.

Once that is accomplished, actually hauling them up is the next challenge. Extending the boom outboard, supported fore and aft with the preventer lines, is usually the most practical technique. Use the main halyard to add support if the topping lift strength is in question. If the main halyard is used to maintain the boom's vertical position, use a jib or spinnaker halyard for the hoist.

Attach a block to the aft boom fitting, and feed the halyard through the block and extend it to the victim. If a separate block and tackle is available, attach that to the boom fitting and connect the line to the victim. Use a primary cockpit winch to hoist the victim back on deck.

I keep a block and tackle in a plastic bag in a cockpit locker on *Voyager* for just these occasions. While I've never had to use it for victim retrieval, it has been very handy for hoisting scuba tanks. It consists of:

½" line.
One double block with cam cleat and becket.
One double block with snap shackle.

Even though a crew might conscientiously practice to be prepared for an overboard situation, the best policy is still to make going overboard impossible. Make strict rules on your boat, and be serious about enforcement so you never have to perform the overboard drill when the chips are actually down.

11

Abandon Ship!

"Make sure you step up into the life raft." When the subject of abandoning ship comes up, that's the gist of what many people know about handling this ominous situation. I believe it's a subject so terrifying to some, they'd rather not think about it, and consequently are unprepared should the eventuality arise. Any procedure that can be your last hope for life is worth consideration and practicing.

When a vessel is jeopardized, a key decision becomes whether to abandon ship for a life raft or to stay with the impaired craft. While conditions on board might have become terrifying and the vessel no longer sailable, it may still provide shelter and present a better target for rescuers than a life raft. Getting off the disabled vessel seems to some a step in the right direction, but the lives of many sailors have been lost in leaving a boat that never sank. This decision should be made with the realization that survival in a life raft is difficult at best, and often becomes the trial of a lifetime for those who manage to step on dry land again.

Though it's a situation we never want to face, the best chances for survival are gained by being prepared for the unthinkable. The choice of a life raft, assembly of the abandon ship supplies, deployment and boarding of the raft, and having a grasp of the techniques necessary to stay alive while adrift at sea are all matters that we need to consider before they actually happen.

The Life Raft

The life raft becomes increasingly important as vessels travel to more remote portions of ocean, farther away from well-traveled sea lanes. It will be expected to house and protect the crew for possibly extended periods of time when rescue is hundreds of miles away. Rafts should be selected on the basis of crew numbers, type of sailing and where it's done, and the quality of the raft's structure and contents. Features of preeminent offshore life rafts include the following:

- independent double bottoms
- insulated floor
- a means of collecting rainwater
- handholds inside and lifelines out
- contains oars ready for use
- inflates within one minute or less
- viewing ports and a quick-opening/closing entrance
- a boarding ladder or ramp and means to pull oneself into the raft
- self-erecting canopy
- abundant ballast for stability
- highly visible exterior (including the bottom)
- muted colors inside
- light inside the raft
- pump mechanism to remove water
- sturdy enough to withstand people landing on it from the vessel
- contains an ample survival pack of quality items
- entryways are closed with zippers or ties, both of which withstand waves better than Velcro. Metal zippers must be cleaned and sprayed with penetrating oil to prevent seizing after each use; nylon is preferred.

Be certain to understand the raft's mechanism of inflation and what a hydrostatic release actually does. Be familiar with the lines used to right the raft should it inflate in an inverted position or be knocked over later. The sea anchor is very important in controlling the raft's drift speed, and enhances raft stability to prevent capsizing and afford a more even ride. Quality of the sea anchor, in size and construction, varies among manufacturers, and becomes a factor worth consideration. Some rafts also provide a second sea anchor.

Ballast systems differ among brands, but all involve water-filled chambers that resist capsizing by wind and waves. Most utilize plastic bags beneath the structure that vary in size and numbers; the larger bags are the most efficient, and are best arranged all around the raft. Most ballast pocket designs permit them to be emptied to improve raft speed when that is desired.

Other designs feature one large ballast chamber and a single bag that encircles the entire raft. A single chamber may provide greater stability against capsizing but could also be disadvantageous in attempting to make the raft sail. The key element is sufficiency of the total volume of ballast water in making the raft stable.

Another concept is that crew weight contributes to raft stability. This makes sense, and says something for sizing the raft correctly for the crew complement. Less crew weight in a given raft translates into the exposure of greater surface area to the elements without a compensatory stabilizing component. Don't buy a raft designed for crews much larger than yours.

Consider also, though, that people crammed into a raft that offers very little space per person would be very uncomfortable, to say the least. I entered a 6-person life raft during a training exercise at a Safety at Sea Seminar along with five other people. Two of the men were very overweight, and probably would not be on an offshore sailing vessel, but the conditions inside were very tight with NO room between people. I shudder at the thought of being wedged tightly in the raft at sea with people that are seasick, possibly injured, scared, and who knows what else.

A polypropylene rescue line with a handle at the end (called a quoit) should be permanently installed on the raft and available for heaving to persons in the water.

The raft must be very sturdily constructed with a high quality of workmanship, able to withstand months adrift, should that become necessary. Neoprene, a synthetic rubber with fiber interwoven for extra strength, is the material of choice for raft construction. The seams of neoprene rafts must be glued together rather than heat sealed, which is a more expensive process but allows for easier repair.

Raft manufacturers should provide full instructions on stowing the raft on board, placing it in the water, inflation techniques and flipping over an inverted raft, boarding, repairs, and maintenance. Video presentations are most helpful. Make certain that no question goes unanswered when obtaining your raft.

A key consideration is where to stow the raft on board your vessel. In the majority of abandon ship situations, there will be time to deploy the raft in a workmanlike manner. The raft has to be instantly available, though, in case the boat is holed severely and begins to sink quickly, leaving precious little time to leave the ship.

Life rafts are packed and protected within soft valises or hard shell plastic or fiberglass containers. Neither container is ever stowed below decks, for obvious reasons. Valises are best located within a cockpit locker, protected from seawater, UV radiation, and being jostled about. Other gear should never cover the valise or make it difficult to pull out. The locker space must be large enough that the raft can't settle and become wedged into a confined area, making it difficult to extricate. Once the valise is set into position, immediately tie the painter, which extends through an aperture, to a sturdy point inside the locker. A line fastened to the underside of the locker lid should be available to hold the lid open.

Hard shell containers are mounted on deck. The mounting cradle, made of stainless steel or anodized aluminum, must be securely fastened to the deck with through bolts and backing plates to prevent its being washed overboard by breaking waves. These are very often attached atop the coach roof beneath the main boom, but the raft containers are not to be used as steps by crewmembers.

Metal straps with turnbuckles and/or webbing are used to hold life raft containers into their cradles. The raft must be held securely in position, but at the same time must be easy to release quickly when necessary. Inspect these routinely for damage to the straps or corrosion that hinders manipulation. Turnbuckle tongs should have ample purchase within their centerpiece, with cotter pins to prevent loosening.

It's also very important to know what survival gear accompanies the life raft, whether it's purchased or rented, and the quality of those items. They could be vital in preserving life, and what the life raft lacks will have to be added to the abandon ship bag. Get a complete list of and examine survival gear items. An example of those included with one offshore life raft follows:

- Bailer
- Canopy lithium lamp
- First aid kit
- Fishing kit
- Floating anchor
- Floating knife
- Food rations
- Graduated cup
- Instruction manual
- Parachute rockets
- Rain catching gear
- Rainwater collecting pouch
- Reflective canopy tape
- Repair kit
- Rescue quoit with line
- Seasickness meds
- Signaling mirror
- Sponges
- Survival instructions
- Thermal blanket
- Water
- Waterproof flashlight
- Whistle

Abandon Ship Bag

The most important components in the ditch bag, in terms of being rescued, are without a doubt the EPIRB, hand-bearing GPS, and satellite phone. The EPIRB emits an internationally-recognized distress signal to aircraft, satellites, land stations, and rescue vessels. These signals are recognized by COSPAS/SARSAT weather and mapping satellites that fix a location and relay it to ground stations located worldwide. These stations transmit data to Mission Control Centers (MCC), where the signals are tracked and the transmitting vessel identified using the database of registered EPIRBs. The MCC then alerts the local Rescue Control Center (RCC). This agency monitors the area of the distress EPIRB and can launch a Search and Rescue team.

The EPIRB should have a battery well within the five-year renewal period, and should be tested frequently to make sure it functions. The entire crew should have instruction on its use.

It's crucial to register your EPIRB with NOAA because your name, telephone number, vessel type and description, and emergency contact information are frequently used in rescue efforts. Registration information and assistance is available by calling 888.212.SAVE or 301.457.5678.

A hand-bearing GPS allows us to communicate our exact position whenever contact is made with rescue agencies. These small units usually operate on AA or AAA batteries, and I always replace them at the onset of an ocean passage and have spares in plastic bags, along with the GPS, in the overboard bag.

Another revelation in our offshore world is the satellite telephone. Making a call from a life raft to transmit your situational information and GPS position need not be any more difficult than placing a call from your own living room. Store the appropriate emergency contact numbers into the phone, and keep a waterproof copy of the numbers with your ship's papers. The phone must be kept dry, in a plastic bag within the abandon ship bag. If the vessel has an external satellite phone antenna, be certain that the portable antenna goes into the raft along with the phone. The antenna should be placed into the abandon ship bag when it is no longer needed ashore. I have had excellent results with Iridium and Inmarsat phones, but poor results with Globalstar. A handheld VHF radio should complete the communications equipment in the abandon ship bag.

As mentioned, items included in the abandon ship bag supplement the life raft's survival kit, and will vary among vessels. The following table illustrates a list of items that should be considered for vessels sailing offshore waters. Items are selected based on crew complement, the waters in which the boat sails, contents of the life raft survival kit, and individual preferences.

The items below are not usually kept in the abandon ship bag while sailing, but should be added before it's taken to the raft:
- satellite messenger
- yacht log
- water jugs tied together
- binoculars
- hand bearing compass
- camera
- portable watermaker/desalinator
- ship's medical kit
- sextant, navigation publications and plotting tools
- waterproof charts of the area
- portable radio
- spare sail or sheet plastic can become sails
- cockpit cushions
- strobe lights (placed on the raft to draw attention).
- extra clothing
- towels (chamois type)
- all-purpose knife
- plastic bags, various sizes
- extra blankets
- gloves

Notice that a portable satellite messenger heads this list. This new technology has changed the prospect of life raft survival as surely as the GPS and satellite telephone. Aside from its continual transmission of the raft's GPS lat/long position via commercial satellite, the capability to send an emergency message to the GEOS International Emergency Response Center, which contacts appropriate rescue agencies all over the world, makes this device a natural and necessary addition to any abandon ship bag. It is standard equipment aboard *Voyager*. See Chapter 1 for a full discussion.

Water in plastic containers should also be taken; milk jugs with screw-on lids are perfect. We lash several containers together at departure for sea with polypropylene line, ready to be taken to the raft with one line should the need arise. The crew should grab as much food as possible, placed in plastic bags to keep it dry. Other items that may prove very useful, especially in remote regions where time before rescue is expected to be prolonged, include:

- The ship's man-overboard pole, which could be used to construct a sail or extend a strobe to attract attention.
- The dinghy (attached with a lanyard).
- Spare sail.

Courtest of Landfall Navigaion

The bag itself should be brightly colored, watertight, and shock resistant (Fig. 11.1). It must be located for easy access, yet out of the way of routine traffic; the salon is usually the best place. If one bag is overfilled or too heavy to carry, the items should be

Fig. 11.1
Voyager's abandon ship bag protects its contents from water and shock and contains items that supplement the life raft's survival pack.

divided into multiple bags. There must be identical overboard bags for each life raft on the vessel.

I use a slip knot to tie the bag to the post beneath the salon table, where it's out of the way yet very accessible. Each crewperson should have a smaller personal bag of items ready in their berths for instant access. This bag contains personal medications, eyeglasses, credit card, cash, hat, warm clothing, camera, sunscreen, and so forth.

The life raft's location and steps involved in its deployment should be discussed with the crew before departure. The abandon ship bag is stocked with most of the implements necessary to complement the life raft survival pack (Fig. 11.2).

-Handheld GPS w/ spare batteries	-Life jackets/survival suits
-EPIRB(s)	-Waterproof VHF
-Satellite phone	-Sat phone portable antenna
-Raft repair items	- Flares
-SART	- Medical kit
-Sun block	- Large black trash bag
-Food	-Water
-Passports	- Can opener
-Hats	-Xyalume® sticks
-Spare batteries	-Portable watermaker
-Hand-bearing compass	-Duct tape
-Crew items	-Spear gun
-Rescue laser	-Dye packets
-Space blanket	-Seasickness medications

Fig. 11.2
Note that the ship's papers are stowed (in a large clear plastic bag) in the overboard bag. All crew passports are gathered, stowed together in a plastic container, and placed in the bag also.

Steven Callahan's *Adrift* is an epic book. Its pages describe one man's struggle to survive for seventy-six days stranded on the Atlantic Ocean in a life raft. I took a lot from this book, but a central theme really hit home: survival depends as much on attitude and determination as on the hardware taken aboard the raft.

We'll discuss procedures for raft deployment, crew boarding, and measures taken while awaiting rescue. What I describe as a "bulldog attitude" is just as essential. Survivors may have to overcome terror, panic, injuries, seasickness, thirst, hunger, and loss of hope; an indefatigable mind-set and faith must overcome all of those obstacles. The average time for rescue at sea worldwide is three days—72 hours after initiation of an EPIRB signal. This time frame is significantly less in more heavily-trafficked waters, and more in remote oceans.

Abandoning ship is one of the emergency situations for which a designated protocol must exist. The whole crew should be familiar with these steps and able to carry them out before going to sea. The following chart depicts the protocols we use and teach before each offshore passage.

Abandon Ship Protocols

* Person detecting the emergency BLOWS THE HORN or calls "EMERGENCY ON DECK" to summon all hands.
* The boat is immediately hove-to.
* All crew members put on lifejackets and foul weather gear/survival suits and life jackets.
* Coordinate efforts to save the boat, depending on the emergency, such as fighting a fire or stemming a serious leak.
  - Make contact with land bases to inform of emergency.
  - Designate two people to deploy the life raft and assure

that the lanyard is tied before its release. They put it into the water in its canister or valise to the leeward side. Pull the painter all the way out of the raft container, and bring the raft close to the vessel for crew boarding.

- The raft is inflated on the captain's order with a hard jerk of the painter line, which remains tied to the vessel.
- One person enters the raft, ready to receive the abandon ship bag, food, water, etc. He or she will also assist the rest of the crew in boarding.
- Once the captain decides the ship is lost, the order is given to abandon ship. All crewmembers then enter the raft.
- Release ditch bag and place it in the raft, along with all other items.
- Activate EPIRB, but only one initially.
- Initiate Mayday distress call or contact land bases to advise that crew is abandoning ship.
- Once all hands are on board, with all gear and supplies needed, assess the immediacy of releasing from the foundered vessel. Stay with the vessel until it begins to go down. At that time, cut the lanyard, paddle away from the sinking vessel, and deploy the sea anchor.

The painter, or static line, connects to a valve on the inflation canister and exits the raft container through a sealed aperture. The bitter end is tied to a solid attachment point after the raft is placed into its storage position on the boat. The painter connects the life raft to the vessel after it's launched into the water, and also triggers raft inflation. It takes up to 15 to 20 pounds of force to actuate the inflation valve. As the raft inflates, bands holding the two halves of a deck container will release, and the container will open and drop away. Pressure from the inflating raft will also force open valise bags to free the raft. Full inflation takes anywhere from a few seconds to a minute or so. A loud hissing sound from the raft when near full inflation comes from the over-inflation release valve; it is normal and should not cause alarm.

There is confusion for some about the function of hydrostatic releases; I consider them to be safety valves in case crewmembers are unable to release the raft before the vessel sinks. The mechanism is activated only after the vessel and raft are 10–15 feet underwater as the vessel sinks. Water pressure activates the hydrostatic mechanism, which then *releases the raft from its cradle*, enabling the raft to float to the surface. The static line pays out as the raft separates from the boat, and eventually becomes tight enough to trigger the CO_2 cylinder to inflate the raft. The hydrostatic mechanism does not directly inflate the raft, nor does it initiate inflation while the vessel remains on the surface.

Attempts to attract attention are paramount whenever vessels are endangered. The International Regulations for Preventing Collision at Sea (COLREGS) recognizes the following distress signals indicating a vessel's need for assistance:

—a gun fired at intervals of one minute
—a continuous sounding of a fog signaling apparatus
—red flares
—SOS Morse code (... _ _ _ ...)
—the word Mayday spoken over the radio telephone
—the international call letters N.C. (November, Charlie)
—a visual signal consisting of a square flag having above or below it a ball or anything resembling a ball
—flames on a vessel
—orange smoke
—slowly and repeatedly raising and lowering arms outstretched to each side

—emergency positioning indicating radio beacon (EPIRB)
—high intensity white light flashing at 50 to 70 times a minute (inland waters only)

Most of these signals can only be effective if other vessels or land is nearby; offshore we must resort to the long range communications of SSB, satellite phone, EPIRB transmissions, and satellite tracker system. Crews should send out Mayday messages with SSB, if time permits, before abandoning the vessel, since the radio will be lost along with the boat. Satellite phones accompany crews into the raft, and can be used after the crew has left the boat. This is one of the reasons that I have opted for a satellite phone for long range communications. Please refer to Chapter 1 for a discussion of long range emergency communications.

A Mayday transmission must be delivered in a calm and clear voice to assure that receivers understand this important message. It contains information that identifies the vessel, position, nature of the emergency, and vessel status.

—Press the transmit button and state very clearly, "Mayday, Mayday, Mayday. This is the vessel _____. My position is _____Latitude by _____Longitude. We are (state the nature of your emergency). We require immediate assistance. We have _____persons on board, (along with whatever other information is important)."
—Listen for a response; re-broadcast, time permitting, if none comes.

The Float Plan (Fig. 11.3, following page) is your backup life-saving mechanism. This important document is filled out after the departure date, itinerary of arrivals/departures to each destination, and crew complement are finalized. It is transmitted to land bases at home and to the ports of landfall if possible; but *not* to the Coast Guard. It is understood by these bases that if communications or arrival at the next destination are more than two days overdue, the Coast Guard is to be alerted of the itinerary and your status.

Voyager
FLOAT PLAN

VESSEL NAME _____

SKIPPER NAME
_____

VESSEL DESCRIPTION:

REGISTRATION NUMBERS
_____

DOCUMENTATION NUMBERS
_____

VESSEL CONTACT NUMBERS
_____

CREW NAMES   **PASSPORT NUMBERS**
_____
_____
_____
_____
_____
_____

VOYAGE ITINERARY

DATE   PORT OF DEPARTURE   DAYS AT SEA   DESTINATION
_____

Courtesy of Further Offshore.

Fig. 11.3
The Float Plan used aboard **Voyager.**

ABANDON SHIP

Should a catastrophic event occur, abandon ship procedures should begin only when the captain deems the emergency situation uncontrollable, and loss of the vessel to be imminent. It is safer to stay on a capsized or foundering vessel than to abandon ship for a life raft (Fig. 11.4 below).

Fig. 11.4
This vessel has taken on water and is in danger of going under. The decision whether to enter the life raft will be made soon.

The boat is more stable, may still provide shelter, and is easier to spot from the air or sea (Fig. 11.5, left) than a life raft. Too many sailors have lost their lives by leaving a vessel that never sank.

Fig. 11.5
A life raft is a small target on a vast ocean.

Releasing the Life Raft

Once that order is given, it should be carried out quickly, with everyone busy with assigned tasks. The boat should be hove-to if possible to alleviate severe motion. A final lat/long position is written down and placed in a waterproof container. Everyone dons foul weather gear or survival suits and life jackets immediately. One or two hands are placed in charge of heaving the raft over the leeward side (the painter is tied securely to the vessel at all times). The raft need not be inflated immediately; it will float within its container until the inflation mechanism is actuated. Bring the raft alongside with the painter line, and leave it floating until the order comes to inflate. A quick, hard pull of the painter inflates the raft.

On occasion, a raft will inflate upside-down or be inverted by the wind right after inflation, and someone has to enter the water to flip it over. Often, it's possible to use the wind to help in righting it. Go to the *leeward* side and grasp the strap that crosses the bottom of the raft. The strap is maneuvered to pass over the raft. Stand or kneel on the edge, then lean back and pull the strap until the raft flips over. If there is no wind, go to the side with the inflation cylinder and pull the strap from there.

One person should board the raft and prepare to receive the abandon ship bag and other supplies, which are handed over from the vessel.

During this time, each crewperson should grab their personal overboard bags, along with the other essentials listed, and prepare to board. Take seasickness medication before entering the raft if time permits. Grab as much food as possible, placed in plastic bags to keep it dry. Items such as the ship's dinghy, a spare sail, MOB pole, blankets, cockpit cushions, etc., are brought aboard the raft as time permits. When the skipper gives that order, the other crewmembers begin boarding the raft.

Crew should board directly from the boat without going into the water if at all possible. It's safer, keeps people dry and warmer, and avoids all of the hazards of being in the water. If time permits, people can attach their tethers to the life raft painter and/or take hold of the raft's safety line to prevent being washed away by wave action. Rafts should be strong enough to allow crewmembers to jump down onto the canopy directly, though that does risk damaging the raft.

If entering the water becomes necessary, lowering yourself slowly minimizes sinking below the surface and can decrease the shock of entering cold water. Always go in feet first to avoid head injury. If jumping in becomes necessary, attempt to do so near the raft. Fill your lungs with air, hold your nose closed, and grip the life preserver to prevent it from pushing up over the head. Avoid landing on another person or flotsam in the water.

Once in the water, move directly toward the raft, grab hold of the safety lines, and get into position to board via the ladder or ramp. Minimize movement in the water; any and all activity causes more rapid cooling of the body temperature than being still. The best position to retain heat is to hold the legs to one's chest, relax and breathe, and let the life jacket provide buoyancy. If several people are in the water, huddling together allows them to help each other stay afloat with less physical effort.

Once all crewmembers are in the raft, the captain climbs aboard. It may be necessary to cut the painter, shove off, and paddle away from the boat immediately if it's sinking, if there's fire aboard or on the surface, or if flotsam threatens to puncture the raft. A foundered sailing vessel makes a far more noticeable target in the water than does a life raft. If the vessel is awash and not sinking, it can pay to keep the raft tethered to it instead of immediately pushing away (Fig. 11.6, page 190). In this situation, the vessel must be observed at all times in case it finally begins to sink. Keep a knife on hand to cut the painter and release the raft from the sinking vessel.

Courtesy U.S. Coast Guard

Fig. 11.6
This foundered vessel is still buoyant and is not sinking. It is much easier to spot from the skies than a life raft.

After cutting the painter, shove away from the boat to avoid the suction of water she can create when slipping below the surface, and put out the sea anchor to stay in the immediate vicinity.

The first few minutes in the raft are a very busy time. The highest priority is to make sure that all survivors are helped out of the water and into the raft. Only re-enter the water if someone is unable to get to the raft without assistance; the rescue line is designed as a throwable device to pull people to the raft. Injuries need to be attended to; those needing more critical care should be tended first, while lesser injuries may have to wait until the situation is stabilized.

Injuries to be confronted at the outset are most likely to include near-drowning, hypothermia, laceration wounds, blunt trauma injuries—including contusions, sprains, and fractures, shock, closed head injuries, and burns. Treatment of these and many other medical problems is discussed in *Further Offshore*.

The EPIRB should be activated if not already done, water bailed out of the raft, and inventory taken of all materials brought aboard.

The sea anchor should be deployed right away to minimize the raft's drift away from the boat. That will be your last known position if a message was broadcast, where search and rescue (SAR) efforts will begin. The sea anchor also lends stability to the raft and dampens its motion.

Examine the raft to become familiar with locations of ports, entryway, oars, and survival pack, canopy light (inside and out), pump, and repair kit. Locate and read instruction manuals for life raft maintenance that may arise.

Be able to open or close canopy ports and the access as conditions indicate. There must always, even during cold or rain, be some ventilation available to avoid carbon dioxide buildup in the air.

Many rafts have a mechanism on the floor to pump out water. Gain an understanding of how this manual pump operates, but be prepared to bail water or use towels to keep the floor as dry as possible, since continued exposure to seawater causes skin sores.

A repair kit will be included within the survival pack. The kit uses glue to hold patches over punctures or tears, but the repairs rely on the opposing segments being dry. Don't waste resources by placing a patch over wet material.

You can also repair a hole by placing a plug within the tear and sealing its edges securely with a clamp. This depends on having adequate fabric on which to place the clamp, so deflating the raft may be necessary. Once enough of the defect's edges are available, secure the clamp very tightly against the plug for the best results. Re-inflation will put tension on the repair edges.

A clamshell clamp may be the best repair option, especially for larger holes or those accessible only from the exterior. One section of the device is placed inside the tear, with its threaded component protruding through the hole. The other half is positioned against the outer hole segment, and the two components are then screwed tightly together. At least one clamshell clamp is recommended for the abandon ship bag since they're not always included in raft survival kits.

Make sure that no sharp objects, whether it's gear or anything worn by the crew, threaten to puncture or abrade the raft. This will be a constant concern, from day one until the day of rescue.

Personal hygiene and raft cleanliness must be addressed. Everyone must agree to urinate and defecate outside the raft when conditions permit. This can be done while tethered to the raft, or in the water if no sharks are around. If rough weather prevents excreting outside, buckets must be available for use inside. These are emptied at the first opportunity.

As food and water intake diminishes over the days to follow, excretions will also occur less frequently, but the need to abide by the guideline is always important. In an inhospitable environment, a clean, sanitary, dry, and orderly raft helps to keep the raft habitable and improves morale of survivors.

Seasickness is very likely to be a factor for most of the crew at this stage. It will not set in initially for many, since adrenalin is an antidote, but once people settle down, the raft's motion, fear, exhaustion, stress, and dehydration will bring it about for some. Medication, if not taken before leaving the ship, should be dosed soon, before vomiting begins. If conditions permit, open the ports and entryway to allow people to see outside rather than seeing just the raft's inner walls. Vomiting should be done outside the raft if possible, but have vomit bags or a bucket readily available to prevent messes all over the raft floor. If someone throws up, be sure to clean it right away, or that victim will soon have plenty of company.

Get the raft warm. Even in the tropics, the sea is still cooler than body temperature, and the raft floor will take on that temperature. Place cushions, sheets of plastic, or blankets over the floor to insulate against the water's chill. The body heat of crewmembers can elevate the temperature inside if heat loss is controlled in cold weather. People huddling together, sharing body heat and decreasing exposure to cold air, is more efficient than being apart. The highest percentage of heat is lost from our heads; wear hats or in some way cover the head to preserve body heat. The onset of frostbite occurs in the extremities when blood is shunted to the body' core. Keep hands, feet, ears, and noses warm with socks, gloves, and blankets, placing hands in the armpits, and lying end to end with a crewmate to warm each other's feet.

Anyone who has spent time in a life raft understands the stark reality; this is where nobody wants to be. The space is very confining; people may be terrified, injured, and seasick. The raft is totally at the mercy of the elements; only a double bottom and its resistance to capsizing prevents disaster. The captain must demonstrate leadership in directing activities and keeping a sense of order among the survivors. Crewmembers respond to assured leadership in a raft just as aboard the boat, and may be in more need of reassurance and confidence than ever before. It is also important to maintain a sense of teamwork and responsibility of all occupants; activities include maintaining the raft, rationing food and water, keeping constant watches, and minimizing conflict. Each person has to understand that their contributions are needed to assure the survival of everyone else, and that the whole crew will get through it together.

Distress messages will begin the search and rescue efforts, which will vary in time frames based on the boat's position. Whether a Mayday was sent out using the SSB, the satellite phone, or by way of an EPIRB, the majority of rescues take place within three days of that broadcast.

We're all familiar with stories of survivors spending weeks adrift in life rafts. In the early stages, set a tone of resource conservation so that nothing is wasted. Become organized by setting watches, discussing the things that have to be accomplished each day, and including all opinions in the conversation. It is important, however, that the skipper makes final decisions based on all input and the thoughtful consideration of factors. If rescue is delayed and time adrift becomes prolonged, an understanding of what will be needed from everyone could be vital to the survival of all involved.

The human body gains heat from the environment by conduction, from warm air by convection, through radiation from the sun and ground, and as a byproduct of internal bodily processes. Though water makes up about 65 percent of its mass, the human body loses water that must be replaced continually. Excess heat must be dissipated through perspiration, resulting in large fluid losses in warm climates. We also lose fluids through respiration, urination, vomiting, and diarrhea.

People suffer from water deprivation much sooner than from a lack of food. Water becomes more important as time in the raft increases. Long term, without inordinate fluid losses, a healthy, active person needs a minimum of one pint of water per day. That requirement increases with excessive losses, so obtaining potable water and using it conservatively often determines survival. Without water, the average person will die within two weeks, and if the weather is very warm or the person is working hard, survival may be limited to only two or three days.

Dehydration is a progressive disorder, with symptoms that reflect its severity (Fig. 11.7 below).

The effects of progressive dehydration:

Loss Of Water (approx.)	Effects
To 4 litres	Thirst, Vague, Discomfort, Impatience, Nausea and loss of Efficiency.
To 8 litres	Dizziness, Headache, Dyspnea, Tingling in the limbs, Decreased Blood Volume, Increased Blood Concentration, Absence of Salivation, Cyanosis, Indistinct Speech and Inability to Walk
To 15 litres	Delirium, Spasticity, Dimming of Vision and Death

Courtesy of Speed Plastics Limited

Fig. 11.7

The long-held belief that people can't survive on seawater has been put into serious question. The Bombard Experience was an experiment done by Dr. Bombard, during which he survived adrift in a raft at sea for 63 days with only the food and water available from the sea. He demonstrated that it is possible to "subsist for more than a month in a decent state in the sea." He determined that humans could survive when ingesting a maximum of 32 ounces of seawater per day, but for a maximum of five days at a time. His recommendation is for crews to begin drinking small amounts (too much causes vomiting) of seawater right away, before their bodies become dehydrated, thus preserving the pure drinking water for later use. The study also determined that thirst is alleviated by immersion in water (presumably not cold water). While the study has merit since it reflects Dr. Bombard's actual survival while drinking seawater, I would leave it to the reader to test his claims in actual practice.

Water in jugs brought aboard the raft will only last so long, after which potable water has to be collected during rain showers or distilled from seawater. Containers should be available so that everyone can help in the effort when rain begins to fall. Water should be stored in sealed containers to prevent leakage and loss.

Water catchment systems on rafts collect rainwater in gutters and then funnel it from the raft's exterior to tubes inside. Salt accumulated on the raft has to be washed off before collection, since it contaminates the first rains to fall on the raft. This foul water can, if collected, be useful for cleansing wounds and cleaning bodies. Clean plastic sheets attached to oars and extended outside the raft can also direct water inside for collection. Two types of commercial devices are available to desalinate seawater; the solar still and reverse-osmosis watermakers.

Stills are fragile, depend on bright sunlight, and can't be used in rough seas. They convert seawater by condensation inside a sealed, inflated balloon. Seawater is introduced from the top, from whence it trickles down through fabric wicks; during this process, sunlight condenses drinkable water as minerals are left behind. The water exits through a hose at the bottom to a collection bottle. Solar stills are usually placed on the sea surface, where the hose and collection bottle are kept cooler.

Portable watermakers are sturdier and more dependable than solar stills. They resemble bicycle tire pumps in that a suction hose extends into the sea, and the device is hand-pumped to create pressure inside the unit. Potable water, created by the process of reverse osmosis, exits from the other end within a few minutes.

Osmosis is the natural movement of solvent from an area of low solute concentration to an area of higher concentration. The process of reverse osmosis used in removing salt from sea water forces a solution of high solvent concentration through a semipermeable membrane to a region of low solute concentration. The pressure required for this conversion is 600–1,000 psi. The membrane forms a barrier that retains solute (salt), and allows water through to a collection tube. In the portable desalinators most suited for use in life rafts, salt rejection is said to be 95.3 percent to 98.4 percent, with an output of about 30 ounces per hour if pressure is maintained. The membrane and O-ring do wear out, so spares should be kept available.

Urine is excreted from the body by the kidneys, which filter the blood of unwanted solutes and minerals. While it is sterile in the absence of infection, and does contain water, re-ingesting the undesirable contents of urine does the body no favors. Only specific filtering, such as that done on the space station, allows the drinking of urine.

While it is not recommended with straight seawater, we can absorb up to a pint of water through the rectal mucosal membranes. This would be most useful if salty rainwater, or water from the raft floor, is available. Fresh water can also be given through enemas;

this can be especially useful in cases where nausea or diminished stomach capacity makes oral digestion difficult. Rubber or plastic hoses, lubricated with ointment and inserted a couple of inches past the anal orifice, work well. Fluid is introduced through the other end with a funnel. While this may be humiliating and embarrassing, it's well worth the effort to maintain hydration.

Some fluid is available in fish; we can suck the liquid from inside of the eyes, and from bones of the spinal vertebrae.

Water will be needed in near-drowning victims or anyone who has ingested a lot of seawater to help filter elevated solute contents in the bloodstream. In all others, hydration levels should be near normal at first, so water intake should be restricted initially.

While we normally consume up to a gallon of fluid daily, humans can survive on as little as 2 to 5 ounces on a daily basis. Begin rationing by dosing at 14 to 16 ounces a day for four days, then try to taper that down as conditions permit. Take water into the mouth to wet the lips, and then slosh it around before swallowing to moisten the oral membranes.

Food is not quite as essential as water. Humans can survive up to a month without food, especially considering the size of many people today. Assess what was rescued from the boat and what was contained in the raft survival pack and abandon ship bag.

Carbohydrates supply energy more readily and require less water for digestion than proteins. Many survival foodstuffs are freeze-dried or dehydrated, and thus can't be eaten without water available, which is also true of such items as crackers, nuts, or pretzels. While they may potentially supply valuable nutrition, if they can't be digested, they're worthless.

It makes sense that body motion requires energy. One tenet of life raft survival is to relax, lie down as much as possible, and conserve energy and fluid resources.

Fish, birds, sea turtles, plankton, and seaweed are the most plentiful sources of food. Fish abound in most oceans, and usually constitute the primary source of food on a life raft. Those available offshore should be free of the toxins found in reef-dwelling species, so that is seldom a problem. They're eaten raw, of course, but that presents no problem besides that of individual palatability, which is quickly overcome by hunger. Everything on a fish can be eaten except the intestines, liver, and roe; even the stomach is OK if flushed of acid first. The bones, too, are edible if chewed thoroughly before swallowing, and they also make good toothpicks and suitable fish hooks.

Raft survivors relate that fish come to dwell beneath life rafts, probably for the shade they provide. Catching fish must become a routine activity, both to provide food and to occupy time in a useful way. Fishing kits provide the gear necessary, except fresh bait. Most fish will bite at lures and whatever pieces of other fish you provide as bait. They're all looking for a meal that won't fight back. Any kind of string or line can be useful, and hooks can be rigged from pieces of metal, wire, plastic, or fish bones.

Always exercise caution when using anything sharp on a raft, because of the danger of puncturing the inflation chambers. Always keep knives, fish hooks, and spear tips covered when not in use, and stow them to prevent accidents. Some fish, such as the Dorado, thrash wildly when hooked, and care must be taken to protect the raft when they're brought aboard.

Spearing is also a useful tactic, especially with fish that "adopt" the raft and stay nearby. Those species that swim slowly around are prime targets; just be careful with the spear gun in your hand and with the tip that protrudes from a fish after a successful hunt.

Fish are attracted by light and avoid the hot sun during daytime, so just as on inland lakes, fishing is often best at night. Flying fish can sometimes be lured into jumping toward a light, and will land on the raft just as they do on a ship's deck at night.

Once a fish is caught, gaffed, speared, or netted, bring it aboard carefully to avoid damaging the raft or your hands. The spines of many species can do real damage, so wear gloves if they're available. Small fish can be hoisted directly with the fishing line; larger catch may slip away if you attempt it. They're best landed by using a makeshift gaff, but you can also scoop them up with your hand or get a grip on the gills.

Once you have it on board, the best way to kill a fish is by severing the spinal cord directly aft of the head. You can also bend the head of smaller fish to sever the cord. Beating a fish over the head with an oar can work, but you run the risk of beating a mate over the head as well or tearing the raft canopy, so be careful. Again, protecting the raft is of ultimate importance, so exercise caution when using sharp objects.

The flesh of freshly caught fish contains fluid, so that's the best time to eat it. All meals should be eaten slowly, bits at a time, to make digestion easier and more complete. Fish are very high in protein, which requires more of our body's water in the digestion process than carbohydrates. Don't overeat a fish meal unless water is plentiful.

It is important to divide portions of any meal equally to prevent crew troubles. There are several methods to assure fairness. If two people are in the raft, one person divides the food and the other picks his/her portion. With more people, a method used by the British Navy for hundreds of years can be employed. After the food is divided as equally as possible, a person closes their eyes and chooses who gets the portion designated at random by another person. This process continues by rotating personnel until all servings have been allotted.

Whatever fish (or any other meal) is left over should be sun dried, if possible. It should be cut into thin strips with fat removed and placed in the direct sunlight. The high humidity of the marine environment can impair the drying process, and the meat can spoil. Whenever the meat takes on a slimy and/or green appearance, it should be discarded.

Birds flying above sailing boats and floating rafts often stop by for a visit, and they can become dinner. If they land on the raft, a quick grab can be successful. A bird once grabbed a fishing lure that a crewman was trailing behind our boat, and that method can also be successful on a raft. Birds will bite at various lures and bait, and you can snag their beaks with a quick jerk of the hook. Others tell of creating a loop with fishing line, placing a bit of a snack in the middle, and snaring birds when they go after the bait.

Birds are best killed by grasping their heads and giving the neck a quick, hard snap to the side to break vertebrae and sever the spinal cord.

Sea turtles also investigate life rafts, and can also be a source of food; the meat, eggs, and blood are extremely nutritious. Whenever other food is available, I would prefer that over turtle meat, since sea turtles are endangered. But since starvation calls for extreme measures, the best way to capture a turtle is by grabbing the hind feet and pulling it into the raft. Be careful of the toenails, which are long and sharp; they can damage your hands and the raft. Many turtles will urinate when handled, so keep that in mind unless you need a shower.

Turtles are slaughtered by severing the carotid arteries on each side of the neck. Most of the edible meat is around the legs; open the turtle by cutting between the top and bottom shells, and then severing ligaments that hold them together in front.

Whether the meal consists of fish, bird, or turtle, any toxins present are likely concentrated in the liver. Discard the liver and intestines, and never eat the eggs of tropical fish, where toxins can also be found. Offal not eaten can always be used as bait to lure in the next meal.

Ciguatera poisoning is caused by toxins in the liver and flesh of reef-dwelling fish derived from their diet of flagellates, coral, and algae. These species are not likely to be present far offshore, and shouldn't be a concern until or unless the raft makes landfall on an island. Avoid eating fish with protruding lower jaws (they eat coral), barracuda or jacks, and never eat fish roe. Turtle eggs *are not* toxic.

Plankton is not present in all waters, but it's usually plentiful in most. These are very small, photosynthetic creatures that drift in the water rather than swimming about; they exist near the surface so they can absorb UV rays. They are nutritious, supplying vitamins (vitamin C prevents scurvy) and protein. Plankton is gathered in nets (the sea anchor is perfect) that can be constructed using any light fabric. Either tow the nets behind the raft (as long as the raft is making way through the water) or skim the water manually with the nets when the raft is drifting. Plankton looks like scum and is smelly, but the taste is not as bad as you might expect.

Seaweed, which is likewise full of vitamins, proteins, carbohydrates, and minerals, also provides bulk. We're all aware of the widespread use of seaweed in Japanese dishes. Seaweed is always salty, and should therefore not be eaten without water. If possible, rinse it with fresh water before eating, unless stores are low. It can be dried in the sun and chewed (as a cow chews its cud) to provide nutrition and stimulate salivation to keep the mouth moist. Sargasso weed is the main variety available offshore in the North Atlantic and Caribbean seas. The berries present in autumn have a somewhat sweet taste and are very good, but the weed itself is tough.

Patches of seaweed always deserve exploration because they often attract small fishes and crabs. These critters can be netted and are eaten whole.

Sharks are known to bump and grind against the bottom of rafts, and that's bad for a number of reasons. First, it scares people, and rightfully so. They may be planning to eat other fish there, or critters such as barnacles attached to the underside. Sharks will also rub against rafts for unknown reasons, and they're also known to bite at things randomly. This can be minimized by keeping the raft inflated to prevent large bulges protruding from the bottom. Sharks have no way of sensing the presence of people in the raft unless something is dangled or swims in the water, so stay inside when sharks are around.

It makes little sense to fish for a shark; the difficulty in landing and killing one without damage to you or the raft is minimal, and is surely a fool's undertaking. Avoid dumping fish guts or any other offal when sharks are in the vicinity; you don't want their curiosity aroused, and you surely don't want to make them frequent visitors.

Encourage sharks to leave the raft alone by striking them on the snout or in the facial area with an oar. They are predators, but their intelligence directs them to look for meals that don't fight back and injure them. Remember that, unless someone from the raft has dangled an appendage or been detected in the water, sharks' prey is fish or barnacles under the raft.

Barnacles are free-floating until they adhere to solid objects in the water, but they wouldn't be expected to appear beneath the raft for several weeks. They can be harvested from the raft's bottom or anything else, and used for food; the whole organism is eaten, including the shell, which contains valuable minerals. Removing barnacles from the raft gives sharks, fish, or sea turtles one less reason to investigate and possibly damage the bottom.

Maintaining the Crew and Raft

Establishing a system of order and team mentality is a primary objective after all hands are aboard the raft. After addressing the concerns prioritized after abandoning ship, focus shifts to chores necessary to maintain the raft, just as a monitoring and maintenance schedule is followed each day aboard the vessel. These duties can be divided into those inside and those outside the raft:

Inside Duties
—raft maintenance, including repair of tears or abrasions
—opening and closing ports as needed
—turning off lights during daylight hours
—pumping/bailing water
—keeping raft inflated
—sanitation
—rationing food and water
—tending to illness/injuries
—navigation
—sustaining morale

Outside Duties
—maintaining a constant watch
—controlling sea anchor as indicated by conditions and navigation
—fishing or obtaining other food
—capturing rain water
—producing water by whatever equipment is present
—preparing food
—raft maintenance
—displaying light at night, turning off during daytime

Medical matters arising in the days or weeks after crewmembers board the life raft are much different than those incurred before and while abandoning ship, usually reflecting the chronic nature of life adrift, and could include:

- —dehydration
- —shock
- —seasickness
- —headache
- —fishhook wounds
- —swimmer's ear
- —constipation
- —diarrhea
- —food poisoning
- —marine animal stings, bites, punctures, or lacerations
- —burns
- —hyperthermic illnesses
- —hypothermic illnesses
- —skin lesions (Sailor's sores, permiosis)

Dehydration is the chief long-term concern, and the constant need to produce or collect water is a daily responsibility. Fluid needs are controlled by minimizing losses, while enough water must be allotted to each individual to sustain body homeostasis and prevent organ failure. The lack of adequate water, and bodily changes that occur, can be related to many other disorders that threaten survivors.

Fluids are transported from cells to the bloodstream when dehydration occurs. This results in decreased ability of those cells to function or outright cell death. This leads to a variety of symptoms. A drop in energy levels, decreased skin turgor (elasticity), dark and concentrated urine, sunken eyes with a lusterless appearance, dry oral mucous membranes, headaches, muscle soreness and weakness, and mental derangement all signify dehydration.

The seasickness that many feel after leaving the ship gradually diminishes for most people, as time allows them to adapt to the sea's motions and to gain control of fear.

Sunburn is always a risk because of prolonged exposure to UV light. Heat-related illnesses result from decreased ability to cool the body; again, dehydration can be an important factor, along with excess physical activity in warm temperatures. Progression of severity ranges from heat exhaustion to heat stroke, which is life threatening.

Hypothermic disorders are also a threat in cooler climates, with frostbite and the three stages of hypothermia all possibilities when adrift in cold waters. Refer to a thorough discussion of these important illnesses in *Further Offshore*.

Scurvy is a disease not discussed frequently since the discovery that ascorbic acid (Vitamin C) is preventative. The onset of symptoms from a chronic dietary lack of Vitamin C takes at least 6-8 weeks. Those fortuitous enough to survive that long adrift would experience sore, bleeding gums and oral mucosal lesions, possibly anemia, with a rapid heart rate and shortness of breath. Bony abnormalities of growth plates occur in children. The symptoms are related in part to defective collagen, one of the body's connective tissues, that is formed when vitamin C is lacking.

Therapy consists of ingesting Vitamin C, none of which is available from a diet of fish and the occasional unfortunate bird. Seaweed and plankton are the best sources.

Heavy Weather

There is much less we can do on a raft than all the measures available to rig for heavy weather aboard ship. We're more defensive, trying to prevent injury and loss with far fewer options available. Wave action and wind are both capable of toppling a life raft, making that our biggest concern in those conditions. The risk of traumatic injury to personnel and the raft, loss of survival items overboard, and flooding inside are all possibilities. The first order of business is to stream the sea anchor (even two if possible). Drag created by an anchor can prevent the surge of wave action from rolling a raft and can provide stability if wind attempts to undermine it from below. Check the sea anchor, its connection to the rode, the anchor line and its associated hardware, and connections to the raft routinely.

Close all ports and entries. Protect the gear and consumables (especially water jugs) against damage or loss. If water jugs and the solar still float outside the raft, bring them inside, along with filleted meat that may be drying on the canopy.

Be sure that nothing sharp can come loose and tear a hole in the raft, and ensure that all sharp points are always kept covered.

Each person should grab and hang tightly onto a handhold. If the raft begins to sway, all crewmembers should counteract that motion with their own to maintain stability. Crew weight should be kept as low as possible to concentrate the center of gravity.

The majority of rough weather experienced at sea lasts from a few minutes to one or two days. Squalls that are produced locally may only bring light rainfall; other types can entail downpouring rain with wind that gusts to 50 or 60 knots. These are all short-lived. Cold fronts associated with low pressure systems can produce thunderstorms, with winds in the 30-knot range that veer to the west-northwest in the northern hemisphere. These are generally not associated with severe weather; winds and seas very often return to prevailing conditions within a couple of days.

The biggest threat is with tropical and extra tropical low pressure systems, which are responsible for the most dangerous storms and hurricanes seen at sea. Certainly anyone sailing the oceans should be aware of the risks attendant upon those waters during hurricane seasons, and assume that risk when opting to be there at those times. Being adrift in a life raft is even less attractive than being aboard a sailing vessel during Force 9+ conditions, and there is no advice beyond that given above for the most severe conditions.

Navigation

The best scenario is without a doubt to have a handheld GPS, with spare batteries, to provide lat/long fixes that can be relayed via satellite phone to facilitate speedy rescue. It is important to keep both instruments dry in a waterproof container, and to have the phone's portable antenna and emergency numbers readily available.

Charts, preferably waterproof, are needed for position plotting, and help in deciding what bodies of land are most accessible. Without GPS, dead reckoning and celestial navigation are the methods available to keep track of the raft's movements. Dead reckoning provides an approximation of position based on the factors of direction, speed, and time where:

Distance in miles = Time in hours X Rate in miles per hour, or D = RT. This formula can also reflect the rate of travel, or speed thus:

R = D/T

This is the formula we use to determine our speed through the water on a raft.

The raft moves primarily by current, secondarily by wind and wave action. Knowledge of the area waters is very helpful here, since current is impossible to detect on the water unless its effects are seen against an immovable object.

Speed through the water (excluding the effects of current) can be estimated by counting the number of feet the raft covers in a given number of seconds, and then translating that into nautical miles per hour. We can utilize some 18th Century methodology, called the *chip log*, to determine feet traveled in a given number of seconds. Tie a strong line to a weighted object so that it holds position in the water as the raft moves away. Tie a knot at the end near the weight, and then tie another knot 16 feet, 8 inches from the first knot. Place the weight into the water next to the raft, and count the seconds that elapse as line is reeled out. Note the total seconds when the second knot is reached.

One nautical mile is 6,072 feet; for purposes of estimation we can round that to 6,000. There are 3,600 seconds in one hour. Based on the formula of R = D/T, we set up the following equation:

$$\text{Speed} = \frac{3600 \text{ s/hr}}{6000 \text{ ft/kt}} \times \frac{\text{Feet}}{\text{Seconds}} = \frac{6}{10} \times \frac{\text{Feet}}{\text{Seconds}}$$

If we selected 16.67 as a constant measure of feet, our computations are made easier:

$$\text{Speed} = .6 \times \frac{16.67}{\text{Seconds}} = \frac{10}{\text{Seconds}}$$

If the raft moved 16.7 feet in 40 seconds, the rate of speed is determined:

$$\text{Knots/hour} = 10/40 = \text{¼ knot.}$$

Over a 24-hour period, the raft would travel 6 nautical miles. Understand that this only gives us the speed *through* the water, since the weighted object will flow with current. Without using a portable GPS, the only way to account for current is by deriving a position fix, based on celestial navigation or by knowing current conditions in the water beforehand. Compare the DR with that position to determine the effects of current.

Direction is calculated by sighting down the sea anchor rode with a hand bearing compass, and then adding or subtracting 180 degrees to arrive at the track line. Recall that wind-generated current is not directly aligned with prevailing wind directions. In the Northern Hemisphere, wind and current are deflected to the right, so expect current to be 15 to 30 degrees right of prevailing wind directions.

Please refer to the many books describing celestial altitudes and sight reduction to obtain LOPs. A sextant, the current *Nautical Almanac*, HO 229 or HO 249, HO 249 Volume I, worksheets, pencils, dividers, plotting sheets, and a parallel rule are tools of the celestial navigation trade.

If you haven't taken the time to learn sight reduction, I would encourage those planning offshore sailing expeditions to do so. However, a celestial computer can reduce sextant altitudes into position fixes, taking the thought and paperwork out of the equation. After obtaining the celestial body's altitude at the precise Greenwich Time, enter the date, time, DR position, body, and altitude into the computer. The computer can generate fixes based on the altitudes of at least two objects taken together or with a running fix of sights taken at least three hours apart. These fixed positions can be plotted on the regular chart or plotting paper.

"Piloting" a raft is far different than doing so on a boat. With no keel, there is no close-hauled sailing toward a destination; rather, the raft pretty much goes with the current, swell, wind, and waves. It's a fool's mission to attempt to travel against a current or waves, a waste of time, muscle power, and mental energy. Make for the largest land mass available located downwind and down current of your best position estimation. Jury rig a sail to take advantage of wind; hoist the sea anchor to cover more miles and deploy it to decrease the effects of adverse winds.

If the raft approaches land, actually making landfall may depend on playing the thermals properly. In most cases wind blows toward islands during the day (sea breeze) and away in the evening (land breeze). In that situation, the anchor is lifted during daytime and deployed at night to ensure progress toward land.

Once a boat sinks below the waves and crewmembers climb aboard the life raft, reaching land or being rescued alive is the ultimate objective. Everything else is done just to stay alive long enough for that to happen. Long range communications at the outset are vitally important to alert home bases or rescue agencies of your situation. Failing that, if your SSB, satellite phone, and EPIRB can't get the message out, survival depends on making it to some land mass or attracting the attention of an airplane or ship close enough to see your signals. The tools useful in gaining the attention of a passing vessel or plane (seen below) should be on the raft either in the survival pack or your abandon ship bag. There's no excuse for coming up short, since it's foolhardy to sail without these items:

—signal mirror to reflect light toward airplanes or ships
—laser light to shine toward airplanes or ships
—flares to attract attention during day or night contacts with planes or ships. The visual distress signals required on a sailing vessel at sea include as a minimum:
—orange flag with black square-and disc (Day); and an S-O-S electric light (Night); or three orange smoke signals, handheld or floating (Day); or three red flares of handheld, meteor, or parachute type (Day/Night)

[Note: most offshore vessels have the three red flares of various configurations. These are usually supplied in flare kits at marine hardware stores. It makes perfect sense to stow the ship's flares in the abandon ship bag where they're available for use on board or in the raft.]

—fluorescene dye packs to attract planes
—rescue streamer
—handheld VHF to speak directly with ships or a rescue helicopter

—binoculars to spot ships, planes, helicopters, or masses
—Xyalume sticks to provide light or attract attention at night
—SART to direct another ship's radar to your position
—strobe light to attract attention at night

These devices must all be close at hand in the abandon ship bag and readily accessible in case another vessel or plane is sighted. They should be kept dry and safe in the bag, which should be near the raft access. Whenever the watch detects a ship or plane, try to attract attention with several crewmembers using more than one method. However, if the vessel fails to see you, do not waste materials on a lost cause.

I have cited the uncomfortable conditions created when people are packed together in a life raft. I can't imagine what it would be like to spend several days under those circumstances. Make sure the life raft is of adequate capacity for the largest crew that may sail on your vessel, instruct your crew on deploying and entering the raft, and do not step into that raft unless it is the last viable option for survival.

Being stranded in a life raft at sea usually proves to be the most difficult trial of people's lives—at least, for those who survive. Our odds are improved with prior consideration of the possibility, which leads to planning and amassing of the equipment and items needed, and developing protocols and procedures for deploying the life raft, bringing the abandon ship bag and other items discussed, and abandoning ship successfully.

From that point on, all efforts are aimed at keeping the raft afloat and people alive until a successful outcome is reached. Develop a team-oriented approach, and maintain a "bulldog attitude" in which success is the only possibility in spite of whatever odds. I know the feeling of pride and accomplishment at the successful landfall of a sailing vessel; I can only imagine the emotions felt by survivors of a life raft cast adrift on the open ocean.

12

Helicopter Evacuation
When and How it's Done

Written by Ed Mapes consulting with Tom Rau

In the 35 years I've spent on the water, I've never encountered a situation calling for emergency helicopter evacuation. That's good news, but it points out an important possible deficiency in my knowledge of how to properly oversee the evacuation of a crewmember from my vessel. When a given procedure is rarely employed, the need to learn and study it becomes greater than one more commonly used. Helicopter evacuation is just such a situation.

The need to summon the Coast Guard for helicopter evacuation of a vessel at sea must be dire. We have no right to occupy the Coast Guard in perilous rescue operations unless lives are truly at stake. Those situations include imminent sinking of the vessel and serious illness or injury of a crewmember that requires rapid professional medical attention. The HH-65C Helicopters (Fig. 12.1) only have a range of about

Fig. 12.1
This HH-65C USCG helicopter is used on extended range search and rescue missions.

300 nautical miles; in other words, the vessel in distress cannot be located more than about 150 miles from the original flight point (i.e. Coast Guard Cutter, Navy ship, oil rig, or land base).

The chopper must have enough fuel to travel to the vessel's location, spend time there in hoisting operations, and then return to base. Vessels farther offshore can only count on helicopter evacuation if a naval vessel is within range for refueling of the chopper (Fig. 12.2, right). Assistance for those boats will most likely come from other ships in the area, merchant or naval.

Fig. 12.2
Refueling operations of a helicopter from a naval vessel.

If within range of a Coast Guard station, effective communications from the vessel in distress are essential. Hailing on VHF channel 16 or 22 will suffice if within about 25 miles of the station. The Coast Guard can also communicate on high frequency single side band radio frequencies from 2,000 kHz to 30,000 kHz. Satellite phones may be the most effective and convenient mode of communication. Please see the Communications chapter for a thorough discussion and list of Coast Guard Regional Centers and the Norfolk Regional Command Center for direct dialing information. After making contact, you will be routed to a Search and Rescue (SAR) Coordinator. This person will then gather information about your situation, whether it involves a sinking vessel or attention for a single crewperson. Be prepared to give information about position and characteristics of the vessel, and if a medical emergency is involved, you'll be asked about the crewperson's medical condition including the nature of the disorder, care already given, how ambulatory they are, and other details. The SAR coordinator will

Helicopter Evacuation

need to learn about your position, weather and sea conditions, and whether abandon ship operations are ongoing. You will most likely be asked to steer toward the helicopter's course to shorten the rescue interval.

If an ill or injured crewperson needs to be evacuated, you may be connected with trained medical personnel who will request the patient's vital medical information. Be prepared to administer first aid and follow any instructions you may be given to control the patient's condition. Maintain a Flow Chart, on which you tabulate the patient's vital signs, treatments given, dosages and times of medications administered, and how the patient responds. If the patient's condition changes, keep the medical contacts updated.

Once the chopper is within hailing distance of the distressed vessel, establish and maintain communications directly with the helicopter pilot on VHF Channel 16.

How evacuees are hoisted into the helicopter depends on wind and sea conditions, the layout of your vessel, and the mobility of the patient. The Coast Guard may in fact send assets other than a helicopter to your location; C-130 cargo planes and Coast Guard cutters are also available as needed. If you are advised that a helicopter is en route, the pilot will initially be concerned with wind and wave action and whether a cable can safely be lowered to your vessel with minimal risk of entanglement with masts, rigging, or deck gear. Extreme weather is beyond your control, and if the elements prevent evacuation directly from the boat, you'll be directed to prepare a life raft or dinghy in which to place the patient. Assuming a direct evacuation from your vessel, your job is to remove all loose objects that could tangle with the cable or rescue device. That includes items such as a bimini top or awning, antenna wires, lines or running rigging, flags, anything hanging from life lines, and cockpit cushions.

Once communications are established with the chopper pilot, advise as to the hoist location on your vessel; this determines how the pilot makes the approach. Shine the searchlight into the sky to aid the pilot in locating your vessel; you may also be asked to launch flares if they would be helpful. Lighting the vessel is extremely important; if the rescue is in restricted visibility, arrange to light up the hoist area with flashlights and a searchlight, and be careful not to shine the lights directly toward the pilot. Once you've been located, the captain may be asked to direct the lighting on the hoisting area or even to extinguish the lighting to facilitate the use of Night Vision Goggles (NVGs).

The chopper's rotors create a downwash that approaches hurricane force winds (Fig. 12.3 right). Everyone on deck must don life jackets, and the pilot will suggest hand signals to use because of the intense noise; even talking on the radio or satellite phone becomes very difficult.

Prepare the patient by putting on a life jacket if possible, and bringing him/her to the designated area on deck. Follow any advice given by medical personnel before evacuation. Place your flow chart of patient information in a plastic bag, and secure that inside the person's clothing.

Fig. 12.3
Helicopter rotor downwash creates hurricane-force winds that can make deck operations more difficult and dangerous.

Courtesy of U.S. Coast Guard

Be prepared to bring and maintain your vessel to a position with the wind approximately 30 degrees off your port bow. If the engine runs, maintain a steady, slow speed straight ahead to minimize boat motion and maintain steerageway. If the engine has failed, you may be asked to heave-to or deploy a sea anchor to keep the bow facing just off the wind. This makes the pilot's job of maintaining a relative position to your boat much easier.

Helicopter Evacuation

Once in position, a rescue basket or litter, with an attached trailing line, will be lowered to your deck from the chopper, as in Fig. 12.4, right.

Do not touch the basket or trailing line until they make contact with the deck, which dissipates a static electrical buildup within the line. Once the trail line or rescue device touches the deck, they will be safe to touch to guide their positioning without receiving a shock (Fig. 12.5, following page).

The pilot will give directions on how to bring the rescue basket on board and load the patient. This is the most desired method of patient recovery because the basket affords a measure of protection for the patient from vessel rigging. It is, however, a more difficult hoist than a rescue litter, mainly due to its larger size. A rescue swimmer will often be lowered to ensure that the patient is properly secured inside.

Fig. 12.4
A rescue basket as it's being lowered from the helicopter.

Designate one crew member to tend the trailing line to prevent it from fouling as the basket is brought on board. Never tie the trail line to the boat; do not couple the basket to the boat in any way. Load the patient into the device and have them keep their hands inside. A hand or arm outside of the basket can be trapped between the device and helicopter.

Load the patient as gently as possible, along with the medical flow sheet. Situate their arms and legs as quickly as

Fig. 12.5
Unique view of the helicopter and rescue basket as it makes contact with this vessel's deck.

possible. If the basket must be taken below to the patient, remove it from the cable and prevent the cable from fouling any part of the vessel. The pilot will hoist the cable away from the boat at that point. Load the patient securely, bring them back on deck, and signal the pilot to lower the cable to deck level. Again, don't touch the cable until it makes contact with the boat. Re-attach the cable to the basket, and signal the pilot that you're ready for the hoist. Steady the basket with the trail line, as seen in Fig. 12.6 (opposite page), with constant tension to steady the movement of the rescue device on its ascent, making the ride much more comfortable and secure for the patient. Do not release the trail line until its full length is paid out and clear of your vessel.

After releasing the trailing line, the hoist is in the hands of the chopper's winch operator and pilot. The vessel should maintain its course and speed until the basket is safely on board the chopper, while maintaining constant communications with the pilot (Fig. 12.7, page 220).

If there are problems bringing the patient to the litter or with the litter itself, a swimmer may be lowered from the helicopter, probably directly into the water, to assist (Fig. 12.8, page 220). Be

Helicopter Evacuation

ready to help the swimmer onto the boat by throwing a line and lowering the swim ladder.

If the weather and sea conditions make direct evacuation from the deck impossible, it may be necessary to prepare a life raft or dinghy. Deploy it to leeward of your vessel, and then tie lines to the craft to keep it alongside. Once it's secured, lower the patient into position. The object is to get it clear of your vessel, so release the attached line so that the dinghy can drift away into a safer hoist location. A swimmer will be lowered from the helicopter to assist the patient into the litter or hoist device. No one from the vessel should enter the water.

Fig. 12.6
This crewperson steadies the rescue basket during its ascent, maintaining constant tension to steady the basket's motion.

Courtesy of U.S. Coast Guard

Helicopter evacuation is in many cases the last line of defense against true catastrophe. The Coast Guard maintains the best equipment available and its personnel is highly trained in rescue operations. It is also the responsibility of sailors at sea to be informed as to the procedures involved and to be capable of assisting in whatever manner is necessary to ensure that rescue operations proceed as smoothly as possible.

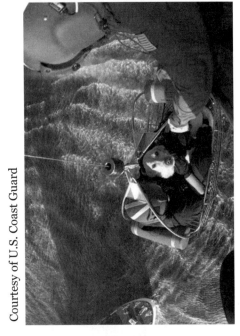

*Fig. 12.7
An important crewmember
is hoisted from a stricken
vessel at sea.*

*Fig. 12.8
A swimmer may be
lowered to assist.*

13

Restricted Visibility

Whether it's a downpour of rain, driving snow, pea soup fog, or the darkness of night, sailing in restricted visibility requires extra precautions. Of the various causes of restricted visibility, sailing on a clear night poses the lowest risk, because we're deprived only of our sense of sight. Whenever water or water vapor in the air restricts our visibility, our ability to hear other vessels or objects on the water is also impaired because sound waves reverberate between droplets and can seem to arise from other than their true source. A fog horn apparently heard from the port bow can actually be on a vessel approaching from the beam; a bell buoy that seems to be at the 3:00 position may in reality be dead ahead.

Fig. 13.1 (following page) illustrates another situation that we should consider as restricted visibility: sailing in big waves. I've watched intently many times as the sidelights of another vessel are visible and then disappear below the waves until our boats are both on a crest again. This, incidentally, is the main reason that I installed a tricolor light on *Voyager*; those lights at the masthead are discernible to another vessel even when my deck running lights are not. I placed my running and stern lights under the control of a toggle switch on the panel, which means they're available in case the tricolor light fails.

Our actions when sailing in reduced visibility, regardless of the reason, are governed by the COLREGS, or International Regulations

Fig. 13.1
Visual contact and even radar images can disappear when large waves intervene in sight lines.

for preventing Collision at Sea. Combining the mandates of the COLREGS with common sense and experience gives mariners the best chance of weathering restricted visibility.

Part Five of this rule states in part that "Every vessel shall at all times maintain a <u>proper lookout by sight and hearing</u> as well as by all available means appropriate in the prevailing circumstances and conditions so as to make a full appraisal of the situation and the risk of collision."

We watch for lights and listen for signs of another vessel or obstruction with the knowledge that what we hear could be deceiving.

Rule 19 is specific for conditions of reduced visibility:

(a) This rule applies to vessels not in sight of one another when navigating in or near an area of restricted visibility.

Restricted Visibility

(b) Every vessel shall proceed at a <u>safe speed</u> adapted to the prevailing circumstances and condition of restricted visibility. A power driven vessel shall have her engines ready for immediate maneuver.

(c) Every vessel shall have due regard to the prevailing circumstances and conditions of restricted visibility when complying with the Rules of Section I of this Part.

(d) A vessel which detects by radar alone the presence of another vessel shall determine if a close-quarters situation is developing and/or risk of collision exists. If so, she shall <u>take avoiding action</u> in ample time, provided that when such action consists of an alteration in course, so far as possible the following shall be avoided:

(i) An alteration of course to port for a vessel forward of the beam, other than for a vessel being overtaken;

(ii) An alteration of course toward a vessel abeam or abaft the beam.

(e) Except where it has been determined that a risk of collision does not exist, every vessel which hears apparently forward of her beam the fog signal of another vessel, or which cannot avoid a close-quarters situation with another vessel forward of her beam, shall reduce her speed to be the minimum at which she can be kept on her course. She shall if necessary take all her way off and in any event navigate with extreme caution until danger of collision is over.

[Recall that when a boat with sails up also uses the motor, it is considered to be power driven.]

(f) In or near an area of restricted visibility, whether by day or night, the signals prescribed in this Rule shall be used as follows:

(i) A power driven vessel making way through the water

shall sound, at intervals of not more than 2 minutes, one prolonged blast.

(ii) A power driven vessel under way but stopped and making no way through the water shall sound at intervals of no more than 2 minutes two prolonged blasts in succession with an interval of about 2 seconds between them.

(c) A vessel not under command, a vessel restricted in her ability to maneuver, a vessel constrained by her draft, **a sailing vessel,** a vessel engaged in fishing and a vessel engaged in towing or pushing another vessel shall, instead of the signals prescribed in paragraph (a) or (b) of this Rule, **sound at intervals of not more than 2 minutes three blasts in succession, namely one prolonged followed by two short blasts.**

Along with maintaining a lookout by sight and sound, slowing down when indicated, and maneuvering to avoid a collision, there are more tools at our disposal: the radio and radar. Whenever visibility is impaired while traversing an area with other traffic, radar is the best means to identify other vessels or obstructions. When radar is available, it should be manned on a continual basis until the conditions change; placing a guard zone around your vessel with an alarm is also warranted.

The VHF radio is another great tool available to keep us out of danger. I contact other vessels at sea on a regular basis, just to make sure I don't dent their hulls, and it makes even more sense to establish communications when the risk of collision, or even a relatively close approach between vessels, exists. This is the practice of good seamanship, and brings about an understanding and agreement of how to proceed in a safe manner.

The radio can also be used to broadcast a securité message on VHF channel 16 to alert other vessels of your position, heading and speed.

The procedure is to broadcast:

1. "Securité, Securité, Securité." (Pronounced see-cure-i-tay)
2. " This is the Sailing Vessel _____."
3. "Our position is _____."
4. "Our course is _____."
5. "We are sailing/motoring in reduced visibility."
6. "Any vessels in this area please respond on Channel 16."

Sailing at Night

It's standard practice to have at least two flashlights and the spotlight in the cockpit at night. Flashlights are used mostly to assess sail trim, and the spotlight is there to shine on the main sail in case another vessel cannot be contacted by radio. These light sources are used sparingly, though, because "night sailing" is only seldom really dark. Don't get me wrong, there's no black like the black of night at sea, when clouds obliterate all lights from the sky. But it's only then that night sailing is truly dark. Even when the moon is a mere sliver or below the horizon, stars at sea cast a surprising amount of light that we can easily sail by. I carry a portable light on a lanyard around my neck each night, but keep the boat darkened to maintain night vision. Once done, turning on a light actually can diminish vision; we see what the light points at, but night vision is lost for several minutes.

Whenever water, in the form of mist, rain, or snow, is in the air, a bright light can be useless, since it's reflected back from the particles in the air. Be very careful, when using a light, not to aim it at a white bulkhead, mast, or even the shrouds; that bright image will imprint on your retina and leave a trace for quite some time.

Shipkeeping is always important while under way, but never more than at night. Keep passageways clear to prevent someone's tripping over an unseen impediment, and stow gear and provisions so that all crewmembers can locate them in dimly-lit conditions.

A chat with the whole crew around dinner time to go over the evening's weather outlook, strategies for sail trim, etc., and what sails are deployed is important. Make sure to reduce sails early if winds are picking up, and be sure everyone knows the next step in sail reduction if it becomes necessary. All hands should know what reef the main is on, what headsail or spinnaker is up, and what halyard is being used. Daylight observations should give a good idea of expected squall action during the night, and procedures for handling these bursts of wind and rain should be understood in advance.

Actions in Fog

Fog is actually a cloud that forms at the surface instead of aloft. It consists of billions of water droplets created when the evaporated moisture at the surface moves upward, cools, and condenses into droplets. It is formed at 100% relative humidity when the air temperature reaches or is below its dew point. Bearing in mind that fog results from warm, relatively moist air that is cooled until it condenses, there are several types of fog that concern mariners.

Frontal Fog

Frontal fog forms when a cold front undermines warm, moist air and forces it to rise over the cooler air. The water vapor cools below its dew point, forming fog that can obliterate objects on the water. Frontal fog is relatively short-lasting, clearing with winds at frontal passage.

Advection Fog

Advection is the process of transporting an air mass to an area that differs in temperature and moisture content. When, for example, warm air moves in from the south, we say it is "advecting" this way. This process can involve moist, warm air over cool water, which occurs most commonly in the spring in areas like the Great Lakes, Chesapeake Bay, and over the Grand Banks of the Atlantic. The warm air mass is cooled by the cold water, condenses, and becomes a thick, unrelenting problem for mariners. Contrarily, cool, dry air can be transported over warm water and the relatively warm and moist air above it. That air is cooled to its dew point, resulting in advection fog seen in the fall months, or in waters near the Gulf Stream. Advection fog often moves in like an opaque mass and blankets an area for extended periods of time until dry wind causes it to dissipate.

Precipitation Fog

When cold rain from clouds high aloft descends toward warmer air near the surface, that air is cooled to form the fog so commonly seen during squalls at sea. Fig. 13.2 (page 229) shows how the horizon is obliterated as grey sea and sky become one. This is commonly seen during squalls at sea, but clears quickly when the hard rain stops.

When fog is forecast and becomes imminent, take steps to prepare the boat and crew to deal with the restricted visibility until it clears. Fog presents one of the most challenging situations for sailors, since it can severely restrict visibility and cause us to misconstrue the direction of sounds.

—Assess the situation, considering what caused the restriction, the proximity to other vessels or obstructions, the distance from destination, and other factors that influence decisions.
—Have the crew don life jackets and use harnesses if on deck.
—Reduce speed if located amongst other traffic or land, including to bare steerageway if necessary.
—Plot a course to remove the vessel from sea lanes as quickly as possible.
—Turn on and monitor radar.
—Use a fog horn.
—Have a handheld VHF radio on deck and ready for use.
—Maneuver to starboard in restricted waterways such as rivers.
—The crew must assist in maintaining watches. Ideally, a crewperson is placed at the bow to watch and listen, but this is impractical during squall or heavy weather conditions.
—Take a position and plot it on a chart before visibility becomes impaired. Steer on a known course and speed to promote dead reckoning in case of electronics failure.

General rules are, when detecting a vessel forward of the beam, (except when overtaking it), to change your course to port, and to steer toward a vessel abeam or abaft the beam of your vessel. These rules are not ironclad, though, and must be undertaken only after thoroughly assessing the situation and the advisability of the chosen maneuver.

If sailing, make certain that the engine is ready to run, or have it on and at idle or switch to auxiliary power altogether for improved maneuverability.

Wind will often cause fog to build up and be more concentrated on the windward side of land masses. With that in mind, one way to escape fog is to get to the leeward side of an island or spit of land, where visibility is often much improved. Another strategy is to get farther away from the windward side by traveling further to sea to get clear of the fog bank concentrated nearer the windward coast.

Sailing in Heavy Weather or Squall Conditions

Fig. 13.2 below illustrates precipitation fog that develops during the hard, cold rains of sea squalls. Heavy weather sailing entails increasing seas with wind-blown spume that also restricts visibility.

Sailing in these conditions requires that the vessel be outfitted with appropriate gear and steps be taken well in advance to best withstand the elements. Please refer to Chapter 14 for a discussion of heavy weather protocols.

Fig. 13.2
The author peers through driving rain during this heavy squall. Note how precipitation fog blurs the land/sea distinction.

Fig. 13.3
This vessel is attempting to heave-to for safety in the midst of storm force winds. Notice the white streaks of foam that develop when wind shears off the tops of waves. That sea foam is also in the air, and combines with spray to restrict visibility.

14

Heavy Weather at Sea

The mere specter of being caught out in storm conditions is the biggest deterrent to ocean sailing for many people. Without actually having experienced a gale at sea, they base their impressions on magazine articles and the exaggerated horror stories of others. It makes day sailors out of many, unfortunately depriving them of some of the greatest sailing days and adventures imaginable.

In point of fact, though, seasoned blue-water sailors recognize deteriorating conditions and cope with them as a matter of course. They understand what forces and circumstances the vessel will be asked to withstand, and then go about the business of making ready in a seamanlike fashion. Not that gales don't make conditions uncomfortable, and pose a risk of crew injury or boat damage, but they do happen and sailors must understand how to cope with heavy weather. They also need to develop confidence in their abilities to do so. For most people, that sense of well-being doesn't come from books or sailing lessons--it comes from experiencing those conditions and witnessing the effectiveness of their actions. That's when the studying, reading, and practice translate into actual seamanship; that's when sailors become truly competent. For me, it was gaining a sense of just how much a well-handled vessel can take; far more than what most people can endure, that's for sure. I know for a certainty that if my boat is well maintained, properly equipped, and appropriate measures are taken when indicated, she will protect me and my crew through virtually anything, and that's very comforting to know.

What is "heavy weather"? The definition will vary greatly between those with limited experience and sailors who have weathered gales at sea, because their preparation, outfitting, actions, and attitudes will be very different. In truth, when voyages are routed according to the most opportune timing for passagemaking, it is actually rare that a crew will have to deal with greater than Force 9 (winds 41-47 kph) conditions (Fig. 14.1, below). I've certainly seen those, most commonly in short-lived squalls or air mass thunderstorms, but only a handful of times since the mid-1980s.

Fig. 14.1
This vessel has its storm stays'l rigged and fights to prevent beam-on waves during Force-10 conditions.

The Beaufort Scale of Winds and Seas defines the effects that wind has on water based on historical data of given wind strengths correlated with the changes they cause in sea conditions. The Scales are very reliable except when those winds blow against a current, such as the Gulf Stream or Aghulhas Current, when sea states can be much worse. Mariners can use the distinctions of the Beaufort Scales to help in their assessment of the conditions they face, in that it helps us understand and categorize the weather to form perceptions of what's to come. The entire Beaufort Scale is located in the Appendix of this book; please see the excerpt in Fig. 14.2.

Beaufort Number	Wind Speed (knots)	Wind Speed (mph)	Sea Wave Height (feet)	Sea Wave Height (meters)	Description	Effects observed on sea
8	34-40	39-46	18.0 - 25.0	5.5 - 7.5	Gale	Moderately high waves of greater length; edges of crests begin to break into spindrift; foam is blown in well-marked streaks.
9	41-47	47-54	23.0 - 32.0	7.0 – 10.0	Strong gale	High waves; sea begins to roll; dense streaks of foam; spray may reduce visibility.
10	48-55	55-63	29.0 - 41.0	9.0 – 12.5	Storm	Very high waves with overlapping crests; sea takes on white appearance as foam is blown in very dense streaks; rolling is heavy and visibility reduced

Courtesy of Further Offshore

Fig. 14.2
This excerpt from the Beaufort Scale of Winds and Seas illustrates various stages (Forces) in wind strength and their equivalent effects on sea state. Notice that each successive Force is associated with more severe wave conditions, calling for corresponding strategic changes in boat handling.

Note that a true "gale"—Force 8—consists of 34–40 knot winds with seas that can reach the 18–25 feet range. A true gale is recognizable because foam blown from wave tops begins to create distinct white streaks on the dark blue water. Waves grow to around

233

20 feet in height, big enough to create deep, uncomfortable rolling motion if taken on the beam. The boat should be maneuvered to take these waves either astern or off the bow (more discussion later).

The National Weather Service also designates 34 knots of wind as the maximum velocity at which able seamen can effectively handle their vessels without great difficulty or utilizing extraordinary measures. The 34-knot radius, as issued by the NWS in storm bulletins, is the distance outward from a storm center at which winds are 34 knots in strength. This is used by mariners in storm avoidance tactics; they plot courses to keep their vessels outside of that radius for optimum survivability. In other words, managing most vessels in winds of 34 knots or more takes us from the realm of sailing a course to that of managing the vessel correctly to optimize safety and survivability. At that point, maintaining the desired course can become secondary to steering the boat to its greatest advantage to cope with conditions.

The uncertainty of predicting storm path behavior is reflected in the 1,2,3 Rule utilized by the NWS (see Fig. 14.3, following page). It states that for every 24-hour period, a zone of safety is created by adding 100 miles to the 34-knot radius in the storm pathway. At 48 hours, the zone becomes 200 miles, and so forth. Sailors located in the storm path use these predictions when deliberating their avoidance tactics.

Never be Surprised at Sea

Fig. 14.3, opposite, is designed to illustrate large scale systematic events at sea. These are basically low pressure centers formed in the tropics or extra-tropically that can be tracked and monitored by meteorologists. The systems are identified on weather charts

HEAVY WEATHER AT SEA

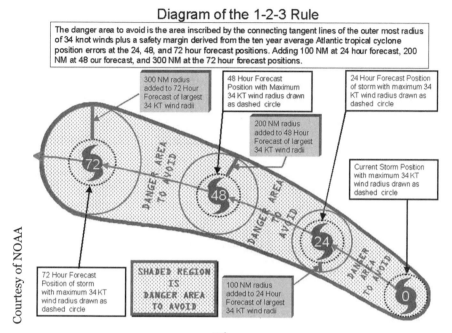

Fig. 14.3
The 34-knot radius *demarcates winds greater than 34 knots. Each day, that distance increases by 100 nautical miles because the storm path becomes more difficult to predict.*

available to us offshore, providing time to alter course, avoid the 34-knot radii of heaviest winds and seas, and keep ourselves safe. Even without downloaded weather information, these systems should not sneak up and surprise sailors who continually observe and log the weather parameters affecting them on a local basis. Maintaining a keen weather eye on local conditions and combining that with weather chart information is an important facet of heavy weather management; never allow yourself to be surprised at sea by the weather.

The crew should contribute to this effort by noting their findings at the conclusion of each watch in the Deck Log (see Appendix for *Voyager*'s Deck Log). This information is then used by the skipper to keep a running tabulation of these parameters,

235

which are combined with weather charts to formulate the most viable forecasts possible. When deterioration is expected, you'll know when, and will have a good idea of how severe it will be and how long it will last.

An important consideration is to observe the formation of clouds and how they progressively change in the local area, particularly those to windward. Be on the alert for cumulus clouds that grow progressively throughout the day to increasing heights. Those showing instability, with sharply-defined lobules at altitude that are visibly in motion, can become cumulonimbus clouds (Fig. 14.4, below), and portend squall action at some point, often during early evening or throughout the night. When these high cumulus clouds build, somebody's getting wet.

Fig. 14.4
Watch throughout the day as cumulonimbus clouds develop: they foretell the need for caution and preparation for the night.

"Normal" squall development implies a single cell, or pulse, thunderstorm. Such a thunderstorm consists of one air updraft and a one-time downdraft. During the towering cumulus, or growing, phase of formation, the updraft of warm air and resulting raindrop formation cease when the weight of water can no longer be suspended aloft. Rain begins to fall downward through the updraft channels, and the drag of air from the falling drops diminishes air updraft. In effect, the falling rain turns the updraft into a downdraft; the supply of rising moist air is cut off and the life of the single cell thunderstorm is curtailed.

These single cell rain squalls develop when updrafts of warm/moist air is localized; under conditions more conducive to instability, larger systems can form in which clusters of numerous individual cells in various stages of maturity coexist and merge. Updrafts, and therefore new cells, continually re-form with rain and high winds following behind. Individual cell updrafts and downdrafts along the line create strong outflow winds moving rapidly ahead of the system. Sometimes such systems will form in a line—this is the *line squall*—that can extend laterally for hundreds of miles; mariners must learn to identify these systems.

They are distinguished by a straight, horizontal, grey line just over the horizon, with the familiar billowing activity aloft indicating areas of extreme instability and severe turbulence. The low-hanging arc of dark grey or green cloudiness is called the *shelf cloud*. This appearance is a result of rain-cooled air spreading out from underneath the squall line, which acts as a mini cold front. The rain-cooled, dense air traveling in front of the system forces warmer surface air upward, where it condenses rapidly at low altitude to create the shelf cloud, as in Fig. 14.5, following page.

There may be several localized thunderstorm cells within the system, each with the ability to produce lightning, strong winds, and cold, stinging rain. These cells are observable on radar, appearing as circular structures within the darker mass of activity. The centers, or eyes, appear white on black-and-white radar screens, and red or yellow on color displays. We are familiar with this appearance from watching radar displays on television weather broadcasts.

Fig. 14.5
Voyager *sails toward a squall line. Notice the white, puffy shelf cloud above the heavy precipitation. This is not a particularly strong system; note the deteriorating cumulus clouds above. The stormsail is about to be hoisted, nonetheless.*

Single cell disturbances are very short-lived, only lasting a few minutes, but can still pack strong wind gusts and sheet rain. Line squalls have the potential to last for hours, as the system continues to develop individual cells. The most violent winds are usually encountered as the wall cloud and cells at the leading edge approach; I've encountered winds up to 60 knots, but the strongest gusts only last a few minutes. The wind gusts usually hit soon after the rain does. By the time you see this progression toward your position, you should have the boat and crew prepared and ready for it.

Squall rain is evident as it approaches, coming toward the boat across the water with heavy downpours that pelt the ocean and obscure the demarcation between water and sky. This rainfall that impacts the ocean surface with great force can actually dampen wind waves on the surface; the ocean is sometimes much calmer when it passes.

So what distinguishes squalls that threaten with high winds and strong downpours of rain from those that produce only a slight wind shift and sprinkles? The most trustworthy indicators of heavy weather potential include:

—Circumscribed lobules within and atop cumulus formations that visibly "boil" at high altitudes.
—The appearance of rain from cloud bases.
—Cloud system extending miles in the distance with its base consisting of a dark, horizontal line just over the horizon.
—White, puffy shelf cloud located just above the base of a line squall.
—Cessation or diminished prevailing winds as the formation approaches.
—Thunder and lightning.
—Radar images that give the impression of being very thick, solid formations instead of appearing thinner, less dense, and wispy. Zones of dense, dark eye wall formations surrounding light-colored or white eyes.
—At night, the thick clouds totally blot out the sky as a solid mass without patches of starlight showing through.

Line squalls bear a close resemblance to approaching cold fronts, which often feature a similar squall line beneath cumulonimbus clouds of rising warm, humid air. Cold fronts are almost always seen on weather charts as portions of larger scale weather systems, whose effects may have already been noted by wind shifts, changes in barometric pressure, and temperature variations. The localized disturbances described above will not appear on weather maps; we must make note of and prepare for them ourselves through our own observations.

Weather that is produced locally, unassociated with large scale meteorological events, is one important consideration, but we must also be aware of and track weather systems that can affect our forecasts. This is done via the long range communications gear (satellite or Single Sideband Radio) of our choice. Weather information is available in many formats from a range of sources. This includes weather charts for our own interpretation, text forecasts, buoy data, and also recommendations from professional routing services. Selection of these options depends on one's comfort levels in interpreting charted and text information; those with the ability to read weather charts often use only that information to arrive at forecasts, and the spectrum continues to those who rely solely on the advice of routing services.

I've been telling students and audiences for years that combining data from multiple weather sources aids in reliable forecasting. I download NOAA charts with the satellite phone while offshore, and we routinely monitor our own weather parameters as well. Conditions that charts predict are sometimes inaccurate, however—especially those extended past the first 48 hours—so actual local conditions must be taken into account to arrive at the best expectations of weather to come. However, another source of *very* reliable weather data is there free for the taking, from sea buoys (Fig. 14.6, opposite page).

Ocean weather buoys are stationed strategically to monitor valuable weather parameters reflecting actual conditions at those locations. This is real information, as recorded and transmitted by the buoys' instruments, that enhances our decision-making at sea. This wealth of information is to be found at the National Data Buoy Center, at: http://www.ndbc.noaa.gov

Fig. 14.6
A sea buoy on station and transmitting valuable weather data.

On that site, locate the zone that encompasses the waters in which you'll sail, and identify buoys there (Fig. 14.7, following page). It makes sense to select sites surrounding your intended course, providing information no matter where the weather is coming from. Each buoy has its own web address, all available on the NDBC site. Just click on the buoy position and its web site appears with a picture of the buoy and the current, live weather information.

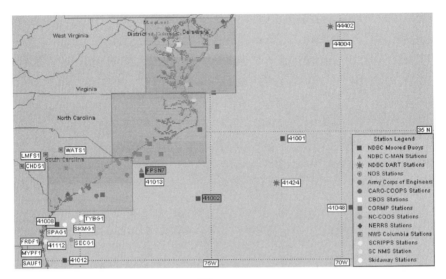

Fig. 14.7
This diagram illustrates sea buoys available for the mid-Atlantic coastal region. Selecting a buoy gets you directly to its website.

Each buoy provides a complete range of weather parameters. This is easily accessed when using land-based Internet. Offshore, though, where download times are an issue, be selective in the information you ask for. I use the following data:

—wind direction
—wind speed
—wave heights
—barometric pressure

I delegate the task of accessing buoys designated for a given passage to a reliable person at a home base. That person logs onto each buoy web site, copies the data for the parameters designated, and then sends the information to us at sea via e-mail. I access the data by simply downloading one e-mail message. Fig. 14.8 is an actual transmission I got during a Bermuda–St. Thomas passage.

```
Hi Ed,
Here is the information from Bermuda to Caribbean as of 10/31 @ 0650.

#41046
wind dir. NE (40 dg. true)
wind speed 23.3 kts
wave ht. 10.5. ft.
atmos. press. 30.13

#41010
wind dir. ENE (60 dg. true)
wind speed 15.5 kts
wave ht. 6.6 ft.
atmos. press. 30.35

#41041
wind dir. E (80 dg. true)
wind speed 15.5 kts.
wave ht. 8.9 ft.
atmos. press. 29.89
```

Fig. 14.8
Actual e-mail transmission from base station gives us valuable weather information on a recent ocean passage.

As you see, Force 5–6 winds were from easterly quadrants, with moderate sailing waves, considering our southerly course. Looks like a great day to be at sea! This is very comforting information, whether it's good or bad weather, because it removes doubt that might exist when local conditions don't always jibe with what we expect from weather charts. The information in Fig. 14.8, above, coincided with a NOAA chart we also studied (Fig. 14.9, page 244).

This indicates a gathering gale to the north and a cold front that approached our track from the west. These systems both pose potential problems to a vessel at 25 degrees N latitude by 064.40 West longitude. The deepening low could shift wind onto our nose out of the south, and a cold front could bring thunderstorms with lightning and stiff winds.

Sea buoys provide mariners with updated, accurate information that can be very helpful in weather forecasting, and that information is free for the taking. I make it a habit to combine data from sea buoys with information gained from downloaded weather charts and from monitoring local weather parameters.

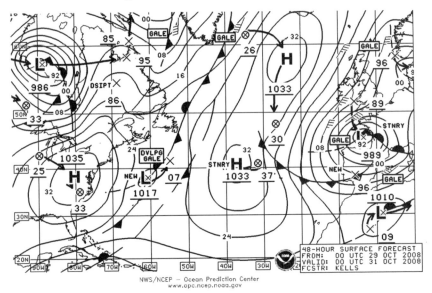

Fig. 14.9
A NOAA weather chart that was combined with the sea buoy information in Fig. 14.8.

Buoys are selected on the basis of the passage route to provide the most beneficial readings. Using buoys at stations that represent directions on all sides of the intended sailing course gives data of weather systems that could approach from anywhere. It is a good idea to have alternate buoys in mind, since it's not uncommon for buoys to be off station or to cease transmissions. In that case, switch to an alternate location that provides information from that quadrant of ocean. Seagoing sailors are known for their resourcefulness; utilizing this great source of weather data is a no-brainer!

Regardless of the outside sources and whatever information is gathered, that alone should not be relied upon to make strategic decisions, since it cannot take locally generated factors into consideration. Only by combining all information available, from both sources, can the most trustworthy and accurate expectation of weather for that vessel be derived.

Avoiding local phenomena can be done by altering course to evade the pathway of those cloud formations. It is even possible to sail through adjacent cells of a multi-cell system by monitoring their radar images and adjusting course as needed.

Whether you detect an organized low pressure system, a cold front, or a localized squall approaching, rigging for heavy weather involves similar steps. This drill becomes routine once you've gone through it a few times, but it makes sense to keep a checklist of tasks handy. This list is also a great topic of conversation with the crew before departure, making them more prepared when the time comes for action.

As I stated earlier, we only rarely encounter severe weather that actually poses a threat to the vessel at sea, and the descriptions above are not intended to instill fear. Experience brings an understanding of the threat posed by squalls and weather systems based on their appearance, how they develop, and whether they're associated with mega-scale disturbances seen on downloaded charts. Localized weather can produce strong winds and heavy rain, though usually of short duration, that can be handled by shortening sail and/or heaving-to temporarily.

It is important to distinguish these phenomena that are seen on a routine basis from large scale weather systems such as depressions, storms, and hurricanes, which pose a much more serious threat.

Rig for Heavy Weather on Deck

Measures taken to ensure safety depend on the layout and equipment of each vessel, but should include actions such as the following:

Reduce to the appropriate sail plan early. A principle of storm sailing is to match your sail plan with the wind strength so that the boat is not over-powered. Since each sailing vessel handles differently, it is important that the master understands the characteristics of his/her boat to reduce sail appropriately. When winds approach and surpass Force 7 (27–33 knots), the main should be double reefed, jib reduced by at least 50 percent, and storm sails made ready for deployment in most cases.

Sail changes should be made early in the game; waiting only makes the job more difficult and even dangerous.

Ensure that a furled genoa cannot deploy by lashing the furling drum into the furled position.

Secure all deck gear so that nothing is loose or flapping in the breeze. Strike the bimini top, which becomes a sail and decreases stability. Never stow a dinghy on a stern davit at sea; they become a serious hazard when large, breaking waves approach from aft. Fill fuel tanks from jerry jugs if necessary. Take cockpit cushions below—unless you intended to replace them anyway!

Secure companionway slides with hatch closed and fastened. There should also be a mechanism ready to secure the hatch from below.

Rig for Heavy Weather Below

—Batten down hatches and dog ports tightly.
—Secure all below-deck latches to prevent spillage.
—Clear passageways to prevent crew tripping over loose articles.
—Pump bilges empty.
—Charge house batteries.
—Close all unnecessary through hulls and seacocks.
—Obtain position fix; communicate position and situation to land base(s).
—Prepare a warm meal, put warm drinks in thermoses.
—Perform the entire routine maintenance and monitoring protocol. (Appendix page 280.)

Crew Preparation

Determine and review storm tactics. The skipper must come to the fore in these situations, because the crew needs confident leadership. There should be a discussion about the steps to follow if weather conditions worsen further. The crewmembers will feel less apprehension when they understand that there are contingency plans and that the boat is secure.

Don and clip on harness/tethers, wear PFD, and carry an overboard alarm transmitter at all times while on deck.

Begin heavy weather watch schedule. While still actively sailing the boat, there should be two crewmembers on deck. Watches are usually shortened to two hours in heavy weather conditions to avoid over-exposure and fatigue. When crewmembers are off watch, it is important that they concentrate on "shutting it off" and trust the other watch to handle the boat. While they're on call for emergency situations, getting good rest is equally important.

Reducing Sail

As stated, sail reduction is always best done earlier instead of later. Reefing the main sail, furling or changing a genoa, and deploying storm sails are all accomplished more easily in moderate rather than strong wind and sea conditions. It is important to consider the seas and not just wind strength; all maneuvers become more difficult and even dangerous when seas are up and boat motion increases.

The following sail reduction guidelines are based on impressions gained from sailing a vast array of boats in varying weather conditions at sea. While they may not be appropriate to every boat out there, the recommendations represent levels that should provide a conservative approach to sail plan management on the vast majority of sailing vessels.

Sloop/Cutter Rig Upwind

(Wind speeds are in true wind.)

15 knots, first reef in main, possibly genoa 10 percent furled.

22 knots, second reef in main, genoa 20 percent furled.

30 knots, third reef in main, genoa 40 percent furled or removed and stowed below.

35 knots, main down traveler deep-reefed or dropped in favor of storm trysail, headsail furled or dropped, storm jib or storm staysail deployed.

40 to 50 knots, employ active close-hauled upwind steering, or heave-to.

50 and above, continue concentrated, active steering or heave-to, with sea anchor if necessary.

Sloop/Cutter Rig Downwind

20 knots, first reef in main. Maintain full genoa.

25 knots, second reef in main, furl genoa 15–20 percent.

30 knots, third reef in main, furl genoa 30 percent.

38 knots, third reef in main or trysail, genoa furled or removed, storm jib or storm staysail deployed.

44 knots, storm trysail and storm jib or staysail.

50 and above, actively sail the waves downwind, deploy drogue to maintain safe speed. Alternatively, head upwind and sail close reached, or heave-to. Never risk sailing too fast downwind under bare poles.

Reducing Sail in Ketch Rig Upwind

16 knots, first reef in main, mizzen full, genoa furled 10 percent.

22 knots, second reef in main, first reef in mizzen, genoa 20 percent furled.

28 knots, main down, single reef in mizzen, genoa 30 percent furled.

35 knots, main down, double reef in mizzen, genoa furled or removed, storm jib or storm staysail (preferred) deployed.

40–45 knots, actively steer close-reached upwind, playing the wind and waves, or heave-to.

50 and above, actively steer or heave-to, deploy sea anchor if necessary.

Ketch Rig Downwind

20 knots, full mainsail, drop mizzen, full genoa.

25 knots, first reef in main, genoa 10 percent furled.

30 knots, second reef in main, genoa 30 percent furled.

35 knots, third reef in main, furl genoa 100 percent, deploy storm jib or storm staysail (storm staysail preferred)

40 knots, drop mainsail, sail on storm jib or storm staysail.

48 knots and above, actively sail under storm staysail, deploy drogue if necessary, head up and close reach, or heave-to.

Courtesy of *Further Offshore*

The basic elements of appropriate sail reduction are main sail reefing, furling in or reducing head sail, and deploying storm sails. It is surprising how many times on a typical passage the main is reefed and then unreefed as conditions change. It is imperative that the crew be competent in this most basic maneuver. Keep in mind that weather changes occur over a twenty-four-hour basis, so we need to able to reef in the dark.

While specific reefing systems vary among boats, the basic steps are:

1) Lower the main sail to within about one foot of the boom.

2) Ease the boom vang, preventer, and main sheet; the boom must be allowed to move upward toward the sail, rather than the sail coming all the way down to the boom. This prevents undue tension on the reef lines, which can cause damage to the lines and the sail.

3) Haul in all reefing lines until the next level is reached. At that point, bring that line in so that the clew is very near the boom. The sail should be pulled aft as well as downward, creating a flattened sail. If the sail is being reefed in less than threatening conditions or down wind, leaving the clew a few inches off the boom will allow for more draft and additional power in the reefed sail.

4) Hoist the main halyard to flatten the sail and move the draft forward. Each successive reef should produce a flatter main sail.

5) Fold the lowered portion of sail and secure against the boom with reefing ties.

Tips for Reefing

Be very careful not to lower the main sail all the way to the boom initially; that causes the reefing line to chafe as it exits the forward boom fitting. Keep the tack cringle/bull ring elevated off the boom to prevent this chafe. Examine the reef lines for chafe frequently: before each passage and daily when en route.

Make sure that the reef lines bring the clew aft and down; pulling the clew aft flattens the sail. Install pad eyes on the sides of the boom in the proper positions, and tie the reef lines through them if necessary.

Coil the reef lines carefully, with a half clockwise twist with each throw to prevent kinks in the line. They should be stowed conveniently for quick use.

Practice reefing and shaking reefs out so that it becomes a routine procedure. Strive to become able to slam in a reef within one minute.

Learn the locations of reefing lines (along with all other lines).

Head Sail Furling

Your furling gear is very important; when a squall approaches at midnight there may not be much time to bring the sail in. The drum and upper furling unit should be inspected routinely, along with furling line and all sheaves. One stout person should be able to furl the headsail by hand under moderate wind conditions. If that is not possible, the mechanism may not be smooth enough.

—The reefing line should be coiled and stowed in a readily accessible location at all times.
—Develop a system for preventing the headsail from unfurling during heavy weather situations. A simple lashing of the drum to the bow pulpit suffices.

Storm Sails

I have gotten into the habit of lashing the storm trysail to the deck above the cabin top prior to passagemaking (Fig. 14.10, following page). This keeps the sail very near its dedicated track on the mast. Whenever breezes freshen to 30 knots, as in Fig. 14.5, the luff slides are placed into the track and halyard is attached to prepare for deployment. The tack lanyard and sheets are permanently installed on the sail.

Whether you use a storm jib or storm staysail, I recommend stowing the sail aft in the boat rather than in the sail locker. Fishing sails out of these forward locations during storm conditions exposes crew to wild bow excursions, soaks them thoroughly, and is not fun. Our storm staysail is usually hanked onto the inner forestay at voyage outset, sheets and tack line in position, needing only to have the halyard hooked on for use. Do not hesitate to deploy storm sails when conditions warrant. Bringing down the main and head sail in favor of sails built to handle gale force winds protects them from excessive wear, and the boat in-

Fig. 14.10
The storm trysail is lashed to the deck and kept ready for deployment.

Fig. 14.11
The storm trysail has been removed from the bag, its luff slides are placed in the mast groove, and the designated halyard and tack line are connected as the wind pipes up.

stantly becomes more stable, less heeled, and more controlled. You can always revert to more sail area when winds subside, but protecting the boat and crew is your primary focus in Force 8 and above.

Making a run for safe haven is very tempting when the forecast calls for storm conditions, and sometimes it's a wise move. Very often, though, it places the boat in greater danger and is the last thing you should do. An important principle when dealing with heavy seas is to avoid a lee shore; only approach shallow water with extreme caution. Maneuvering in large following seas as the boat enters shallower water can be very difficult if not impossible as the kinetic energy of breaking waves takes control of the boat. A loss of sailing ability or of engine power, when needed to crawl off a lee shore or reef, could spell disaster (Fig. 14.12, below) for those who would risk making for port in the wrong circumstances.

Fig. 14.12
This unfortunate vessel was trapped against a lee shore in storm conditions.

While wave heights may grow during a storm at sea, their periods are generally regular, becoming widely spaced in deep water, and they'll break only occasionally if at all. When boats are handled with seamanlike tactics, the greatest margin for safety is in deep water with plenty of sea room to maneuver. Most modern ocean sailing vessels are built to cope with high winds and larger waves in relative safety and comfort. Gaining this understanding and the confidence it brings is fundamental and essential to coping with heavy weather situations.

We only see the upper portion of ocean waves; they extend downward and rotate in a circular motion as the entire wave proceeds along its course (Fig. 14.13, below).

Wave propagation

A = Deep water. The <u>orbital</u> motion of fluid particles.

B = At shallow water. The <u>elliptical</u> movement of a fluid particle flattens with decreasing depth.

Fig. 14.13
Normal wave propagation

When the bottom of the wave is disrupted by a reef, shoal, or opposing tidal current, energy is transmitted to the free-flowing portion of the wave. This translates into more rapid motion and breaking of the crest (Fig. 14.14, opposite), creating hazardous sailing conditions.

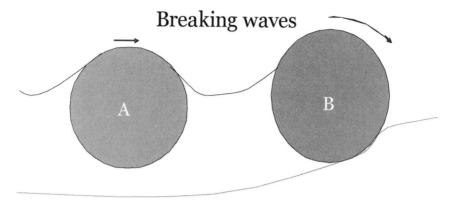

Shoaling effect:
(1) wave height increases

(2) waves **break** as the crest moves faster than the trough and the base can no longer support its top, causing it to collapse forward

Fig. 14.14
Waves become more dangerous as they approach shallow water, moving more rapidly and breaking as the bottom "trips" the lower portion of the wave.

However, there are a few scenarios that warrant making a run for shelter:

—The vessel is in imminent danger of foundering or sinking.

—A crewmember is seriously ill or injured.

—There is ample time to make port and secure the vessel before the storm.

—You have local knowledge, or charts depict a harbor that is safe to enter despite wind, wave, tide, shoals, and other factors that can spell disaster.

Avoid Intimidation

The ocean can be a scary place when it shows its darker side. Foreknowledge of impending heavy weather should afford enough time to make adequate preparations and to think clearly about what must be done. The crew is dependent upon sure-handed leadership whenever placed in unfamiliar situations, and the skipper must rise to the fore as the confident, knowledgeable manager of the vessel.

One of the first changes noted when winds approach Force 8 is the elevation in noise level; loud voices are needed in the cockpit just to communicate. This contributes to anxiety needlessly, because the noise is harmless.

Wind whistles through rigging, sails flutter, waves can roar as they pass harmlessly under the keel, all making for a scary scene. Concentrate on the important factors: wind strength along with wave height and direction. Noise is not a factor. Make your preparations and adjustments as a matter of routine and the boat will handle the rest; trust in that truth.

Gale sailing at night (nightfall occurs every night at sea too!) is another circumstance that most crews would rather not think about. The darkness has a way of making waves bigger, winds more terrific, and danger more imminent. Make certain that sails are shortened and all precautions taken heading into a night at sea. This includes chatting with the crew to review tactics, answer questions and concerns, and address any defects or equipment malfunctions and make any repairs necessary. This relieves crew anxiety, keeps the boat under control, and allows you to get some rest.

Specific Storm Tactics

Heaving-To

The fundamental and most important tactic for all sailors to master is heaving-to. This maneuver can be used so often, in such a variety of situations, that there should be a law requiring mastery of the technique as a prerequisite for boat ownership. Here are some common indications for the use of this indispensable tactic:

—When you near landfall but can't enter the harbor during nighttime hours—never a safe practice without intimate harbor knowledge. The wisest course is to spend the night calmly hove-to, basking in the glow of having made landfall.

—When making repairs. The boat is often more stable and the job made easier when hove to.

—When a crew member is injured. Heaving-to allows full attention for the victim without concern for the vessel.

—If fire breaks out or if the boat is holed, stopping the forward motion is often necessary.

—During a crew overboard situation, especially when sailing downwind, the first maneuver can be heaving-to. This stops the boat's progress away from the victim, allows you to throw flotation aids, and lets you prepare to maneuver for the pick-up.

—When a squall is imminent, slam in a heave-to, take the crew below out of the fury, enjoy a sandwich, and then crack off to resume sailing when the winds subside (Fig. 14.15, following page).

In storm conditions, when actively steering the boat becomes too much, heaving-to is the first tactic used by most mariners.

*Fig. 14.15
View from below decks while the boat is hove to during a particularly determined squall.
The coffee tastes so good when rain pelts down on deck and you're comfortable and dry below.*

Lift is no longer created by the sails while lying hove-to; rather there is only the foresail's tendency to push the bow to leeward and the rudder and/or main sail's effort to head more to windward. The boat is hove-to by tacking, preferably from port to starboard, without releasing the head sail sheet, back winding that sail to drive the bow off the wind. The main sail traveler position is controlled so that the main sail forces the bow toward the wind, creating a balance with the head sail. The mainsail and helm positions are locked in to maintain a close-hauled attitude to the wind and seas. A wind angle of 45 degrees to about 60 degrees off the bow is ideal. The boat will head up slightly and then bear off, but she'll maintain a safe attitude to the elements while making leeway.

The leeway is crucial in high wave conditions, since oncoming waves are dampened by the "slick" of smooth water created from the hull's lateral movement through the water.

In this model, "head sail" can refer to anything from a full genoa to a storm staysail; "main sail" refers to either a full or reefed main or storm trysail, whatever is deployed.

Whatever the sail combination, the main sail square footage should be roughly equivalent to that of head sail deployed.

Experience has taught me that if anything, the foresail can be a bit less, and as flat as possible. Some boats actually heave-to best without a main up at all, or with various helm positions. This has to be determined by experience on that boat. (Refer to Fig. 14.1)

Once the boat is stabilized, seas are dampened as the boat makes leeway, and the motion below becomes much more tolerable. The crew is able to get below, take in some food, and rest as the boat takes control.

The main factors to monitor while lying hove-to include preventing exposure of the beams to the seas and for chafe of sails and rigging. The main area of chafe is between the shrouds and headsail sheet (Fig. 14.16, right). In most cases, flying smaller head sails requires different sheet lead positions, often inside of the shrouds instead of the position of the outboard placement of larger working sails. It pays to know the sheet lead positions of all head sails on the boat well in advance of their deployment; never allow jib sheets to chafe on shrouds when hove-to for longer than a short duration squall.

Wave action from abeam is not only unsafe and extremely uncomfortable, but catastrophic rollovers usually occur when a wave tips the boat past its righting capabilities when driven to leeward. This is true no matter what the size or configuration of vessel; if a wave

Fig. 14.16
The sheet of this 50 percent furled jib lies against the shroud when Voyager *is hove-to for a passing squall. The lazy sheet would be led inboard to heave-to for longer than a few minutes.*

reaches a height of 60 percent of the LOA, it will capsize the boat. If the conditions start altering the boat's configuration, pull the main higher on the traveler, sheet the head sail in tighter to flatten it, or steer higher with the helm.

The boat can be expected to make at least a knot per hour of leeway while keeping headway to a minimum. Heaving-to over hours or even days does require ample sea room to leeward. A distinct advantage of any procedure that keeps the bow into the storm is that the foul weather passes far more quickly than when a boat runs with the storm.

Forereaching

Many skippers, myself included, prefer to manage the boat *actively* rather than trusting in *passive* tactics in which the boat is left to her own devices. There are limits to what people can physically and mentally withstand, but there are great advantages to hand-steering the boat. The most important edge that we have over any self-steering system or the boat itself is the ability to anticipate bigger waves or wind gusts before they arrive, giving us time to adjust course as needed. I know that a motivated crew can keep this up for extended periods of time to keep the boat safe; I've done just that for almost four days in storm conditions at sea.

Another fundamental heavy weather tactic, forereaching, involves steering the vessel with the bows at an optimum 45 to 60 degrees to the elements, protecting the stern and companionway from pooping waves. Driving the boat, even in big waves as depicted in Fig. 14.17, opposite page, is not difficult when she's reefed properly. The object is to ascend the wave faces and then head up slightly toward the crest, where exposure to wind is maximal, and then bear off slightly down the back side to avoid plunging headlong into the bottom.

With some time and practice on the wheel, it is relatively easy to develop a feel for the boat's heel angles as she descends the waves:

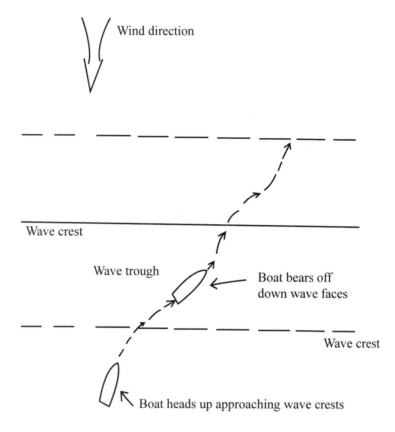

Fig. 14.17
Forereaching in heavy seas. The vessel is steered up (into the wind) on wave crests and down a few degrees as it descends the back side of a wave toward the trough.

not too steeply downward yet not too shallow either, to prevent falling laterally off the wave.

It is even possible to set the autopilot to hold a course, and I've actually been able to balance the sails properly so that the boat sails on her own while forereaching—while I observe from a position near the wheel, of course.

If the crew becomes unable to continue manning the helm while forereaching, a simple progression would be to heave-to, keeping the bows facing into the elements.

Sea Anchor

If and when to deploy a sea anchor will always be controversial, with viewpoints ranging from those who use them whenever winds approach Force 8 to others unwilling to have them on board at all. Advanced technology in modern sea anchors means newer hardware and materials that improve performance. They are still, however, basically parachutes deployed to windward, attached to a stout line that holds a boat's bow into the elements (Fig. 14.18, below).

Fig. 14.18
Sea anchor deployed off its bridle in smooth water during a training exercise.

Survivability is enhanced on most monohulls by two factors.
1) Maintaining the optimum angle to wind and seas.
2) Preventing slack in the rode by sustaining constant tension. Zack Smith, of Fiorentino, asserts that "Rode tension is what I feel is the 'Big Secret' in successfully using a parachute sea anchor."

Just as in lying hove-to, most monohulls riding a sea anchor fare best when the angle of attack is from 45 to 60 degrees to the elements. Fig. 14.19, following page, illustrates the components of a modern, high technology sea anchor as it holds the bow in proper alignment with wind and waves.

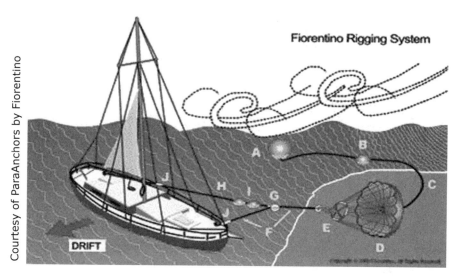

A. Retrieval float.
B. Trip line support.
C. Trip line.
D. Parachute anchor.
E. Pararing, to which the parachute lines attach, and which is itself swiveled to the anchor rode.
F. Anchor rode.
G. Snatch block.
H. Pendant line for bridle control.
I. Support floats for pendant line.
J. Chafe gear.

Fig. 14.19
A modern sea anchor and its components.

The pendant line attaches to a snatch block placed on the primary rode. It extends to the cockpit through a turning block and then leads to a primary winch for easy control. Tensioning the pendant pulls the stern to windward so that the bow heads off; easing the pendant allows the bow to move to windward (Fig. 14.20). Easing the pendant line brings the bow up, closer to the wind and waves, and prevents exposure of the beams to wave action.

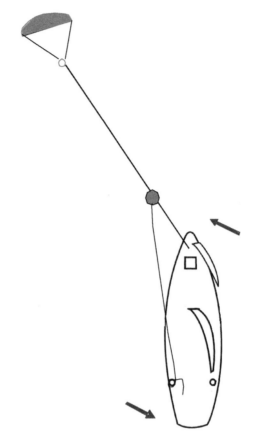

Fig. 14.20
Here, the control line has been eased. Note the bow moving toward the seas when the stern falls off to leeward.

There are vessels on which the sea anchor can't be bridled; some center cockpit designs or a boat with the primary winches located on top of the companionway area are examples. Multihulls are best situated with the sea anchor off the bow, bridled from both hulls to spread shock loading between multiple points of attachment.

The other key factor in lying to a sea anchor is in maintaining optimal rode tension. Rode naturally stretches under force until it becomes taut. As force is reduced, the line relaxes. We want to avoid long periods and amplitudes of slack rode because this leaves a vessel swinging beam-to, where waves can heavily roll the boat or in more rare circumstance, cause it to fall back on the rudder(s).

Limit the amount of rode deployed initially to avoid excesses. Even though you may have the standard length of sea anchor rode available (Vessel LOA multiplied by 10), that much is only meant for use in survival conditions, where waves are very large and spaced great distances apart. Of the 460 feet of rode on *Voyager*, I only deploy about 250 feet initially in 35 knots of wind. After the boat settles in, we reassess how she's riding and look for chafe between all lines and points of contact. Excessive rode allows it to slacken intermittently since the respective positions of boat and anchor are in dynamic flux.

Using a sea anchor oversized for the boat will create more rode tension. The biggest down side to that strategy is in the difficulty of retrieving the larger sea anchor. Says Zack Smith, "I've conducted several on-the-water classes with married couples that plan offshore cruising. Not to sound sexist, but the more petite ladies couldn't retrieve any para-anchor that was over 18 feet in diameter, and that was under calm conditions." His research indicates to me that when it comes to equipment, greater size is apparently not always better. I recommend sticking with the appropriate size equipment, possibly moving up one category if this is the skipper's primary tactic for heavy weather management.

Some other tricks to improve rode tension include:

—Adding a small length of chain next to the parachute. This maintains a more constant resistance by keeping it submerged beneath the surface.

—Purchasing rode with limited stretchiness. Discuss this option with your sea anchor manufacturer.

—Flying a riding sail to increase vessel windage to create a more even "pull" against the anchor line.

While this can be an adjunct on virtually any vessel, the riding sail is especially useful for a boat that can't use a bridle system. The riding sail (Figs. 14.21 & 14.22, below), flown like a

Courtesy of Fiorentino Para-Anchor

Fig. 14.21 *Fig. 14.22*

The riding sail is fast and easy to set up. Use a halyard to secure the top of the triangle sail. The port and starboard sides can be cleated on both stern quarters or tied to the pushpit.

spinnaker, should be set above your sailboat's dodger so wind flow is not interrupted. The concept of this formula is to capture more wind that pushes your boat backwards to create extra tension on the anchor line. Jess Gregory Banner Bay Marine, LLC (www.bannerbaymarine.com) is a noted West Coast authority of riding sails. His rule of thumb for sail size is to take the length of a boat, say 40 feet, subtract 10, and use that as a starting figure in square feet (30 square feet). If the boat really swings a lot, make it a bit bigger.

Next, determine where your riding sail will be rigged. If on the backstay, that's good: it is far back for the best leverage, and the backstay offers a firm purchase. However, if you don't have a backstay, you might either string it up from the stern, tensioned with your main halyard, or attach it to the boom. Under this kind of tension, the sail can sway. If you attach it to your boom, it is further forward, so in either case, you should make it bigger, to gain the same leverage.

The standard material used is heavy Dacron, say 8-ounce, well treated for UV resistance, cut flat, and pulled flat. However, the common complaint is that the sail slats and makes noise as the boat swings and tacks. Using a tough but "soft" cloth such as 12-ounce "Top Gun" actually works better. It is practically silent, and it resists UV rays and mildew well. (Most riding sails are packed away damp from overnight dew.)

A new style sail, the FinDelta, is about 25 percent better than the traditional sail. This claim is based on a NASA-sponsored fluid dynamics study, as well as field testing. The sail has a forward-facing fin, with dihedral wings extending to port and starboard. The sail "centers" the boat earlier in the swing before momentum builds. Talk to your sailmaker or a specialist like Jess to determine your riding sail needs.

I've found that it makes the most sense to get all of the hardware involved directly from the sea anchor manufacturer, both from a cost efficiency standpoint and to ensure that it's correctly sized for your boat.

Deploying a Sea Anchor
Connect the anchor rode to the sea anchor swivel.
Lead the rode fairly (beneath life lines, bow pulpit, etc.) to a secure attachment point on the windward bow.
Place the bridle snatch block onto the sea anchor rode.
Run the bridle line to a cockpit primary winch on the windward quarter.
Release the sea anchor from the windward bow and allow the line to pay out as the boat drifts to leeward.
Pay out line in relation to the wind strength; only use the full rode (10 feet per foot of LOA) in true survival conditions, i.e. Force 10 or above.
Ease the bridle tension to bring the bow closer to the wind;

tension to pull the stern closer to the wind and allow the bow to fall off.

Wherever anchor line touches the hull, install chafe guards. Chafe is the chief hazard of lying-to a sea anchor.

When at an optimum angle, the boat should make leeway with very little forward motion. This creates a slick of smoother water to windward that dampens the wave action coming toward the boat. Monitor the leeward vs. forward movement, adjusting as necessary to maintain the correct configuration to the elements.

On monohulls that cannot bridle the sea anchor to the windward side and on multihulls, the rode is deployed straight off the bow.

Sea Anchor Retrieval

The helmsperson motors slowly into approaching waves toward the retrieval float located at the tail end of the trip line. Meanwhile, the deck crew hauls in slackened deployment line and stacks it on deck or inside a stowage bag. *Do not allow the vessel to overrun the rode.* As soon as the retrieval float is next to the boat, cleat the deployment rode and grab the retrieval float or any portion of the yellow trip line with a boat hook. Haul in on the trip line to collapse the parachute, making it much easier to retrieve.

It is essential that the sea anchor rode be flaked properly after retrieval. Make certain that the parachute portion is clear of its shrouds and can deploy freely at its next use. Use fresh water to remove seawater from the canopy, shrouds, and especially from metallic components such as the swivel and snatch block. I always spray them with penetrating oil after rinsing with fresh water to prevent the onset of corrosion and ensure smooth operation the next time.

Heaving-to is my preferred passive management technique, used most often during limited-duration squalls. If while forereaching during more prolonged storm conditions I feel the need to get below, I do it by heaving-to. Should wind and seas begin to force the bow off to leeward and the boat start to sail forward past its slick, deploying the sea anchor as an adjunct is my next tactic.

Running with the Storm

This was the most-favored technique used to weather storms on square-riggers and clipper ships, and is still employed on modern sailing vessels. Running in the general direction of heavy weather, in many cases, allows us to benefit from the wind and waves as our vessels make great progress toward a destination. These advantages don't come without risks, though, and the crew that runs with a storm must account for each of these risks before sailing directly into serious trouble.

Needless to say, running requires a lot of sea room, possibly taking the vessel hundreds of miles off course. I would be far less inclined to run in that circumstance, preferring to forereach or heave-to to stay closer to my original course line.

Decreased apparent wind velocities experienced while running can be very deceptive, especially in the 25-knot to 40-knot range of true wind velocities and while winds are still building. Apparent winds that feel 6–12 knots less than true make a considerable difference, and the dangers posed by the true wind become very easy to underestimate. Watch *true* wind velocities carefully, and pay attention to the seas using the Beaufort Scale of Winds and Seas as a guide. Think about the difficulty of driving the vessel with the storm, and in turning into the elements to heave-to or set a sea anchor if that should become necessary.

Consider the fact that when sailing along with the system, the boat will be immersed in heavy weather conditions that test the endurance of vessel and crew much longer than would forereaching or heaving-to. Be very careful in running under greatly reduced canvas, a sure indication that wind has picked up a lot. Never run under bare poles; you'll have nothing up to maneuver the boat to heave-to, and excessive speed is a real threat.

The greatest hazard attendant with running is of gaining excessive speed and the difficulties in steering that result. The possibility of broaching or driving the bow into a wave, with devastating consequences, becomes very real when helmspersons are unaccustomed to those speeds.

Speed can be controlled by setting a drogue or objects fastened to the end of a long warp (Fig. 14.23), or rigging a long bight of stout line aft. The objective is to create drag that slows the boat when needed most, while heading down the back side of big waves.

The warp can be comprised of virtually any long line, or lines tied together. Good options include:

- anchor line(s)
- sea anchor rode
- halyards
- spinnaker sheets/guys
- jib sheets

Always rig a bridle that disperses the pull around the vessel, rather than on a single attachment point on the transom. Lead forward to turning blocks along the deck, and then back to cockpit winches.

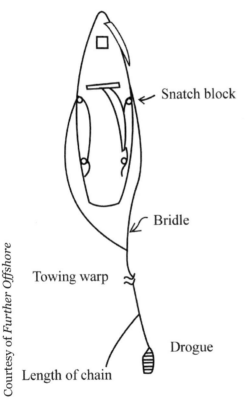

Fig. 14.23
This diagram illustrates the bridle used to deploy a warp aft.

A chain attached near the distant end keeps the warp submerged and pulling. Resistance is increased by tying a series of knots in the line or by attaching a variety of common shipboard articles such as a drogue, cushions, fenders, small sails, the collision mat, sail bags, and so forth.

My friend Hans Bernwell, of Scanmar International, relates that his life was saved by an automobile tire tied to a long warp to control speed as he ran with a storm while sailing off the coast of South Africa. It would have been "suicide," he says, for their small boat to heave-to against those monstrous, breaking waves; they ran with the storm and survived to tell the story.

People have to actively drive the boat when waves build to 20, 30, 40 feet and beyond; no self-steering system can be trusted to keep her out of serious trouble (Fig. 14.24, below).

Fig. 14.24
This 25-footer posed no problem: we steered down a few degrees; it missed our quarter and broke just as it approached the transom.

Qualified helmsmen see, hear, and feel the wind. Even more importantly, they see the wave action, and their anticipation of its approach permits maneuvering to avoid being rolled or having the stern spun around when a big one slams into the quarter. Steering the vessel under these circumstances requires sustained, concentrated efforts by all helmspersons. This, combined with seasickness, reduced food intake, and lack of sleep can eventually cause fatigue that can lead to that one moment when inattentiveness results in a mistake at the wheel.

Handling the wheel is done much like sailing the waves while forereaching: heading up at the crests and bearing off down the back sides. Running fast in big seas and winds is far more demanding, though, and requires more skill to maneuver the boat safely.

Helmspersons must gain an understanding of heading higher on the crests, and then steering down the big waves at just the right angle. Too shallow and the boat can trip onto its beam; too deep and the speed can bury the bow into the trough.

Our crew of four was divided into teams of two during Force 9 storm conditions several years ago while sailing between New York and Bermuda, and we'd just concluded the third day in its grip. The autopilot had crapped out at the onset of rough weather, so hand steering was the order of the day. We took turns driving in ½-hour stints as the watch mate sat harnessed in the cockpit. We said little; the sound of wind whistling through rigging and sails, the water as it roared past, made communication difficult.

Waves from the northwest approached from the port quarter; the wind had now veered northerly. Though meals had consisted of Granola bars, crackers, and bits of apple, we were strangely not that hungry. I'm sure seasickness had a lot to do with that.

Sleep came with difficulty the first two days, lying in the dank bunk with full foul weather gear and harness still on, the noise of the boat slicing through water and the unrelenting, shrieking wind that was impossible to ignore. Now, though, exhaustion had made sleep easy to come by—difficult, in fact, to ward off. I was fitfully sleeping in the starboard aft berth after the midnight–0400 watch. I had succumbed almost immediately after reaching "my world" in the quarter berth, managing to shut it off, to gain some ground on the fatigue that had overtaken me.

One becomes accustomed to the boat's sounds and motions, using those as parameters against which to compare anything different. The degree of the ship's roll was the first clue that nudged me into semi-consciousness. The violence with which we hit the water the next instant was at first startling, and then terrifying as I realized that I was lying against the inner wall of the berth. I struggled to gain access to the small port looking into the cockpit, turned on the small light I always have on a lanyard around my neck, and stared, unbelieving, into nothing but water. The boat was on her side, mast in the water, cockpit filled, and two mates were screaming in terror on deck!

A wave broke over the boat and I felt her lurch to leeward, and then slowly begin to stand up against the onslaught. I climbed/crawled through the companionway and slid the hatch open. The boat was still deeply heeled as the ocean licked at the coach roof, but made slow progress in righting herself. The mast was now above the waves and water poured back into the sea through a long tear in the deep-reefed main sail. Terror was on full display through the saucer-like eyes of my drenched mates as they both hung on to whatever for their very lives. These can be the consequences of driving the back side of big waves, for just a moment, too shallow.

Lying A-hull

This procedure is best left out of your bag of tricks. It involves striking all sail down to bare poles, fixing the rudder to windward, and trusting the boat to do the rest. The only times that I ever lie a-hull is to service the boat in mild conditions or ride out a short-lived squall *in moderate seas only*. The boat naturally assumes a beam-to broad reaching configuration, taking the seas where it's most vulnerable. Seas from near the beam cause a very uncomfortable rolling motion sure to concern the crew, and place the boat in danger of a rollover. Never lie a-hull in big or building seas.

While survival of heavy weather is the stuff of countless sea stories, our efforts should be to keep our own stories boring by avoiding big winds and seas whenever possible. This begins with sensible route planning and continues with vigilance each day at sea. When heavy weather can't be avoided, though, knowing how to handle it will keep the boat and crew safe.

Appendix

Tool Kit

Here's a suggestion for the seagoing tool kit. Sailors should use this as a guide to formulating their own.

- Complete supply of fasteners including screws, nuts, and washers.
- Duct tape, of course.
- Assortment of screwdrivers, both flat head and Phillips head, including tiny screwdrivers for repair of eyeglasses.
- Pliers
- Needle nose pliers
- Socket wrench kit, standard and metric
- Wire cutters
- Wire strippers
- Soldering kit
- Grommet making kit
- Rivet gun
- Large bolt cutters
- Sturdy set of Vise grips
- Crescent wrenches
- Adjustable wrenches
- Monkey wrenches
- Allen wrench kit
- Grips to remove oil filters
- Hand brace and bits
- Hand drill and bits, rechargeable drill with inverter on board
- Socket set including metric sockets
- Hacksaw with spare blades
- Box cutters with spare blades

- Ruler
- Small wood saw with spare blades
- Clamps and/or a vise
- Sandpaper
- Tin snips
- Hammer
- Chisels
- Electrician's snake wire
- Multimeter
- Winch repair kits
- Wire strippers
- Crimper tool
- Soldering kit
- Head repair kits including all necessary replacement parts

My first book, *Ready to Sail*, was a comprehensive examination of the vessel inspection system I've used to ensure seaworthiness of 64 offshore deliveries and all of my offshore teaching voyages. This thorough inspection of the boat has prevented numerous mishaps that would have transpired later had the defects (often minor when discovered at the dock) worsened in the harsh elements encountered on the high seas.

In spite of our rigorous attempts to forestall untoward events, gear does break or cease to function, and we must affect repairs or jury rig a temporary fix in order to make safe port.

We've discussed the need for tools that enable our work; now follows a listing of suggested spare parts that will also be essential.

Fiberglass Repair
- Resin
- Resin catalyst
- Fiberglass cloth
- Roving
- Tape
- Winch repair kits

Rigging
- Sail slides
- Hanks, if appropriate
- Grommet kit
- Turnbuckles
- Toggles
- Cotter pins
- Clevis pins
- Shackles
- Chain or straps to lengthen standing rigging
- Cable clamps, nicropress unit, Sta-Lock or other system for fabricating terminal eyes
- One strand of rigging wire equal in length to the longest stay on board
- Wire cutters sturdy enough to cut all rigging on board
- Electrician's snake wire (reefing line replacement)

Steering System
- Rudderhead fitting
- Replacement parts kit for electronic autopilot
- Replacement parts kit for wind vane
- Preconceived plan for rudder replacement, with portions of the rig fabricated and ready for use

Engine and Transmission
- Belts
- Hoses
- Hose clamps
- Engine oil, at least two gallons
- Transmission fluid
- Water pump
- Water pump impellers
- Starter solenoid
- Starter motor, except if a crank mechanism is available
- Alternator
- Fuel pump
- Fuel filters, primary and secondary
- Antifreeze, if in colder climates
- Injector lines
- DRIP-FREE™ or other stuffing box repair material
- Penetrating oil
- Gasket paper
- Engine gasket kit
- Gearbox oil seal
- Gasket compound
- Gas engine spare parts: points, condenser, rotor, impeller and pin, diaphragm, ignition spray, spark plugs, coil

Electrical System
- Batteries of all sizes of appliances on board
- Bulbs of all appropriate sizes
- Fuses
- Electrical tape
- Copper wire of various sizes
- Wire strippers
- Ring and captive-fork wire terminals
- Crimper tool
- Soldering kit

Plumbing
- Head repair kit including all necessary replacement parts
- Sanitary hose replacements
- Hose clamps
- Hose connectors
- Silicone
- A siphon hose
- Wood bungs
- Two 50-foot non-garden drinking-water-safe hoses and water filter with garden hose connectors. Hoses that store on a reel are preferred.

Sail Repair
- Sailor's palm
- Seamstick tape, 1/2"
- Sewing needles, assorted sizes, straight and curved
- Seizing wire, stainless, 1/16"
- Adhesive Dacron tape, 2" x 2 oz., 3" x 3.8 oz., 6" x 8 oz.
- Tubular webbing, strapping 1"
- Spinnaker repair tape, 2" x 25'
- UV resistant thread, V92
- Waxed nylon thread
- Stainless steel shackles, 3/4", 1"
- Leechline cleats, aluminum
- Batten pocket elastic, 1-1/2", 3" x 8 oz., 6" x 8 oz.
- Dacron tape: 3" x 3.9 oz., 6" x 8 oz., 2" x 5 oz.
- Replacement slides, slugs, hanks, protectors
- Leech line Dacron, #505, 1/8", Kevlar, Spectra
- Shears, bent blade
- Awl, 2-1/2"
- Seam rippers
- Wire cutters
- Hot knife
- Spinnaker "dots"
- Seamstik glue or glue stick

Monitoring and Maintenance Checklist

Task	Schedule
State of battery charge	At least twice daily
Engine	At least twice daily, or as needed
Belts, hoses, leaking, fluids, oil, transmission fluid, coolant, stuffing box.	Twice daily, more as needed
Raw water intake strainer	As indicated
Fuel reserves	At least daily, or as dictated
Water reserves	Daily
Wet-cell battery fluid levels	At least weekly
Bilges	At least six times daily
Propane reserves	At least weekly
Food reserves	At least weekly
Lights	At least once daily
Running rigging	At least twice daily
Standing rigging	At least twice daily
Sails	At least twice daily

APPENDIX

Voyager's Deck Log

Deck Log

Date	Time	Track	Bear	Wind Dir	Wind Spd	Wave Dir	Wave Ht	Swell Dir	Swell Ht	Baro	Clouds	Batts	Bilge

This is an example of the deck log that crewmembers fill out at the conclusion of their watch. It is designed to assist in monitoring conditions on a continual basis.

Contact Information for Contributors

Travis Blain, Mack Sails
3129 SE Dominica Terrace
Stuart, FL 34997
Phone: 772-283-2306
FAX: 772-283-2433
Toll Free: 800-428-1384

Speed Plastics Limited
Wheatbridge Road
Chesterfield
Derbyshire S40 2AB, UK
Tel: +44 (0)1246 276510
Fax: +44 (0)1246 245400
http://www.speedplastics.co.uk/

Blue Water Marine Engines
30201 South River Road
Harrison Township, MI 48045
Tel: 586-468-6960
Fax: 586-468-7737
http://www.blue-water-marine.com/

Rig-Rite Inc.
Phone: (001) 401-739-1140
FAX: (001) 401-739-1149
www.RigRite.com

National Oceanic and Atmospheric Administration (NOAA)
www.noaa.gov

Jess Gregory, Banner Bay Marine, LLC
633 Arlington Avenue
Westfield, New Jersey 07090
Phone: 201 452-2834
www.bannerbaymarine.com

Landfall Navigation
151 Harvard Avenue
Stamford, CT 06902
800-941-2219
www.LandfallNavigation.com

Tom Rau, Author
Boat Smart Chronicles
www.boatsmart.net/book/

Mike Meer
Southbound Cruising Services LLC
7416 Edgewood Rd.
Annapolis, MD 21403
(410) 626-6060
www.yelp.com/biz/southbound-
   cruising-services-llc-annapolis

United States Coast Guard

National Data Buoy Center
www.ndbc.noaa.gov

Zack Smith
ParaAnchors by Fiorentino
1048 Irvine Ave., #489
Newport Beach, CA 92660
(800) 777-0732 (U.S. Only)
or (949) 631-2336
FAX: (949) 722-0454
http://www.paraanchor.com/

Appendix

Beaufort Scale of Winds and Seas

Beaufort number	Wind Speed [knots]	Wind Speed [mph]	Wind Speed [m/s]	Sea Wave Height [feet]	Sea Wave Height [meters]	WMO Description	Effects observed on sea
0	under 1	under 1	0.0 - 0.2	~0	~0	Calm	Sea like mirror.
1	1 - 3	1 - 3	0.3 - 1.5	0.25	0.1	Light air	Ripples with appearance of scales; no foam crests.
2	4 - 6	4 - 7	1.6 - 3.3	0.5 - 1.0	0.2 - 0.3	Light breeze	Small wavelets; crests of glassy appearance, not breaking.
3	7 - 10	8 - 12	3.4 - 5.4	2.0 - 3.0	0.6 - 1.0	Gentle breeze	Large wavelets; crests begin to break; scattered whitecaps.
4	11 - 16	13 - 18	5.5 - 7.9	3.5 - 5.0	1.0 - 1.5	Moderate breeze	Small waves, becoming longer; numerous whitecaps.
5	17 - 21	19 - 24	8.0 - 10.7	6.0 - 8.0	2 - 2.5	Fresh breeze	Moderate waves, taking longer form; many whitecaps; some spray.
6	22 - 27	25 - 31	10.8 - 13.8	9.5 - 13.0	3.0 - 4.0	Strong breeze	Larger waves forming; whitecaps everywhere; more spray.
7	28 - 33	32 - 38	13.9 - 17.1	13.5 - 19.0	4.0 - 5.5	Near gale	Sea heaps up; white foam from breaking waves begin to be blown in streaks.
8	34 - 40	39 - 46	17.2 - 20.7	18.0 - 25.0	5.5 - 7.5	Gale	Moderately high waves of greater length; edges of crests begin to break into spindrift; foam is blown in well-marked streaks.
9	41 - 47	47 - 54	20.8 - 24.4	23.0 - 32.0	7.0 - 10.0	Strong gale	High waves; sea begins to roll; dense streaks of foam; spray may reduce visibility.
10	48 - 55	55 - 63	24.5 - 28.4	29.0 - 41.0	9.0 - 12.5	Storm	Very high waves with overhanging crests; sea takes white appearance as foam is blown in very dense streaks; rolling is heavy and visibility reduced.
11	56 - 63	64 - 72	28.5 - 32.6	37.0 - 52.0	11.5 - 16.0	Violent storm	Exceptionally high waves; sea covered with white foam patches; visibility still more reduced.
12	64 and over	73 and over	32.7 and over	45 and over	14 and over	Hurricane	Air filled with foam; sea completely white with driving spray; visibility greatly reduced.

Index

Symbols
1,2,3 Rule 234
34-knot radius 234, 235

A
abandon ship iv, 3, 88, 146, 153, 173–212
  bag 3, 178–181
    components 178, 179
  decision to, 187
  float plan 185, 186
  protocols 182, 183
Adrift 182
alternator brackets 43
anchor line 22, 26, 28, 40, 118, 206, 265, 266, 268, 270
anchor well 143
Atlantic Area Command Center 9
autopilot ix, x, 44, 53, 99, 103, 108, 109, 136, 162, 261, 272, 277
  CPU 109, 110
  display 109
  fluxgate compass 109
  sensor arm 110

B
Baja filter 48
batteries 23, 34, 41, 44, 45, 89, 157, 178, 207, 246
battery terminals 148
Beaufort Scale iv, 232, 233, 269, 283

bilge 15, 138
  pump 140
  wires 148
blockage 43
block and tackle 79, 90, 113, 172
boathook 27, 143
Bombard Experience 195
bottom watch 14
bungs 137, 138, 145

C
cable clamps 82, 277
CastLok 83
catenary 28
CE 115, 116
Center of Effort 115
Center of Lateral Resistance 115
Central Command 9
chafe guards 126
chafing 31, 67, 73, 78, 81, 126, 148
chainplates 64
chip log, 207
ciguatera poisoning 201
Clamp-Tite 79
clamshell clamp 191
clouds 225, 227, 236, 238, 239, 240
CLR 115, 116
CNG 151
Coast Guard Regional Centers 214
Coast Pilot 14
collision mat 119, 143, 144, 145, 271

COLREGS 15, 184, 221, 222
communications iv, 1, 7, 214
computer 4
   mounts 4
contaminants 52
contributors' contact info 282
corrosion 4, 47, 64, 66, 67, 68, 69, 71, 102, 105, 106, 109, 110, 122, 148, 177, 268
crew overboard iv, 159–172
   equipment
      block and tackle 172
      jacklines 160, 161
      life ring 163
      LifeSling 167, 170, 171
      overboard alarm 161
      spotlight 168
   practice 168
   procedure 162
   protocols 159
   victim retrieval 162–172
      deep beam reach 165
      quick stop technique 164
      Rescue Search pattern 169

D

damaged sail 127
dead reckoning 207, 228
deck log 235, 281
deep beam reach 165, 166
dehydration 194, 204
depth gauge 14
depth sounder 24, 109
derelict 38
diesel fuel algae 47
Digital Selective Calling 6, 8
dismasting 86

distress frequencies 7, 8
   Marine SSB 8
   SSB 7
   U.S. Coast Guard contact numbers 9
distress message 12
distress signal 6, 178
District Command Center 9, 10
ditch bag (*see* abandon ship bag)
dorade vents 136
drag 33, 117, 118, 119, 120, 237, 270
DSC 6, 7, 8
duct tape 128, 275

E

electrical fires 147, 148
electrical wiring 147
electrician's wire fish 75
electronic navigation 4
e-mail 1, 2, 242, 243
emergency contact numbers
   District Command 10
   U.S. Coast Guard 9
      Atlantic Area 9
      Central Command 9
      Marine Safety Center 9
      Norfolk Command 9
Emergency DCC Contact 10
emergency tiller 100, 108, 143
engine 41-60
   air entry 58
   belts 43
   failure 47
   fouled fuel filter 50, 51, 52
   impeller 55, 56, 57
   maintenance checklist 43, 44

engine (cont'd)
    oil levels 58
    O-ring 55
    overheating 53–58
    spare parts 46
    thermostat 57, 78
engine room fires 149
EPIRB vii, 6, 178, 182, 183, 185, 190, 193, 210
external antenna 2
external solenoid 152
extinguishers 138, 139, 149, 152, 153–157
extraordinary towage 37

F
FCC identification number 7
figure of eight ix
FinDelta 267
Fiorentino v, 262, 263, 266
fire iv, 147–158
    causes
        flammable liquids 151
        gasoline 150
        overheating 149
        shore power 148
        wiring 147
            chafe 148
            corrosion 148
            undersized wires 148
    extinguishers 153–157
    prevention
        correct stove usage 151
        propane sniffer 151
        protocols 152
    firefighting 153
    flammable liquids 151

float plan 185, 186
flooding 135–146
    bilge 133
        pump 140
    bow 142
    breach 139
    collision mat 143, 144
    intake hose 145
flow chart 215
fluorescene dye packs 210
fluxgate 109
fog horn 221, 228
forereaching 260, 261
frontal fog 226
fuel
    additive 49
    filters 43, 52
        fouled 50, 51, 52
    quality 49
    system 43, 47
fuel line obstruction 43, 49, 50
furling 251
Further Offshore 88, 125, 131, 134, 139, 162, 164, 166, 167, 169, 186, 205, 233, 249, 270, 292

G
gale vii, 231, 233, 243, 251
galvanic isolator 106
gasoline fires 149
gasoline odor 150
gearshift 44
gelcoat 15
GEOS International Emergency Response Center 180
gin pole 90, 91, 94
grounding 13–24

guy lines 95

H

handheld GPS 1, 44, 45, 89, 207
handheld VHF 10, 11, 89, 179, 210, 228
harbor charts 13
harnesses 159
heaving line 27
heaving-to 257
heavy weather iv, 205, 229, 231–274
   1,2,3 Rule 234, 235
   34-knot radius 234
   Beaufort Scale 232, 233, 234
   clouds 236
      cumulus 236
      cumulonimbus 236
      shelf cloud 237
   crew preparation 247
   deck log 235, 281
   definition 232
   information 240
      NOAA chart 243, 244
      sea buoys 240, 241, 242
   lee shore 253
   noise 256
   reducing sail 247–249
   reefing 249, 250
   rigging for 245, 246
   squall 236, 239
      line squall 237, 238
   storm sails 251
   storm tactics
      forereaching 260, 261
      heaving-to 257–260
      running with the storm 269–272

heavy weather storm tactics (cont'd)
   sea anchor 262
      deploying 267, 268
      lying a-hull 273
      retrieval 268
      riding sail 266
      rode tension 265
      true wind velocities 269
helicopter evacuation iv, 213–220
   communications 214
   hoist location 215
   lights 216
   range 213, 214
   rescue basket 217
      loading 218, 219
      swimmer assist 218, 219, 220
hull diagram 138, 144, 145
hydraulic fluid 103
hydrogen sulfate 49
hydrostatic release 175, 184

I

icicle hitch 84, 92, 97
impeller 55, 56, 57
Inland Navigational Rules of 1980 34
Inmarsat 2, 179
in phase 30
inspection checklist 62
Iridium 2, 179

J

jacklines 160
jib halyards 75
jury rigging a rudder 122

K

kedge 22, 24
keel bolts 15, 20

L

lanyards 34, 143, 163
laptop computer 1
lap zone 106
lateral resistance 116
leaks 137, 144
leeway 117, 258, 259, 260, 268
life jackets 89, 92, 143, 182, 188, 216, 228
life raft 1, 3, 12, 88, 146, 173, 174–177, 178, 179, 180, 181, 182, 183, 187, 189, 190, 191, 193, 197, 198, 203, 205, 206, 209, 211, 212, 215, 219
    ballast systems 175
    boarding 189
    construction 176
    containers 176, 177
    hydrostatic releases 184
    inflation 183, 184, 188
    maintenance 191, 202
    Mayday messages 185
    navigation 207–209
    painter 183
    release 188
    sanitation 191, 192
    stowing 176
    survival 12, 179, 180, 181, 197
        dehydration 194
        food sources 197–202
        gear 177
        heavy weather 205, 206

life raft survival (cont'd)
    water sources 194–197
life ring 163
LifeSling 167, 170, 171
litter 217, 218, 219
LPGs 151
luff slides 126
luff tapes 126
lying a-hull 27

M

mainsail headboard 126
mainsheet traveler system 79
manual bilge pump 51, 138, 140, 142
marine peril 36, 38
Marine Safety Center 9
Marine SSB distress frequencies 8
maritime lien 39
Maritime Mobile Service Identity 7
mast attachment points 64
mast failure 69, 70, 71, 72
masthead sheave 73, 75
mast retrieval 91
Mayday 183, 184, 185, 193
MCC 178
misalignment 68
Mission Control Centers 178
MMSI number 7, 12
monel 71
monitoring and maintenance checklist 280
mounting cradle 177

N

National Data Buoy Center v, 241
National Weather Service 234

Appendix

navigation v, 207, 282, 292
navigational buoys 14
NDBC 241
neoprene 176
night vision 225
night vision goggles 216
no cure, no pay 39
noise level 256
Norfolk Command 9, 214
Norseman 83
NWS 234

O

obstructed fuel lines 50
obstructed primary fuel filter 50
oil levels 58
O-ring 55, 196
osmosis 196
overboard alarm 159
overheating 53, 149

P

Pacific Area Command Center 9
paper charts 45
pendulum 111, 112, 113, 114
pendulum lines 112
penetrating oil 47, 174, 268
portable watermaker 179, 196
precipitation fog 227
preventer lines 79
propane sniffer 151

Q

quick stop technique 164
quoit 176

R

radial unit 101
RAM mounts 4
raw water intake filter 24
raw water strainer 43
Ready to Sail vii, 46, 62, 63, 100, 182, 276, 292
reduced visibility 221–230
    COLREGS rules 221–224
        signals 224
    radar use 224
    sailing at night 225, 226
    in fog 226, 227, 228
    in heavy weather 229
    VHF radio 224
reef cringles 126
reefing 63, 116, 247, 250
refrigeration compressor 43
rescue 21 5, 6
    basket 217, 218, 219
    line 176, 190
    search pattern 169
    streamer 210
riding sail 265, 266, 267
rigging 61–98
    dismasting 86–98
    failures iv, 61
    halyard wrap 76
    inspection checklist 63
    jib halyards 75
    jury rig 92–98
    life expectancy 64
    mast failure 69, 72
    mast retrieval 90–97
    motion fatigue 65
    spreaders 71
    static loading 64

swage fittings 66, 67
ring replacement 133
R (cont'd)
rode 262, 264
roving fender 33
rudder
   control 100
   head 102, 103, 104, 105
   jury rigging 122
   loss 114
   post 101, 102, 103, 104, 105, 106, 108, 122
   stops 103, 104, 107
   tube 102, 106, 108
running aground iv, 13
running with the storm 269
rust 64, 68, 157

S

sail maintenance 127
sail plan 97, 115, 117, 245, 247
sail reduction guidelines 247
sail repair iv, 125–134, 279
   dots 133
   glue 132
   kit 134
   thread 131
   tubular webbing 132, 133
salvage 15, 20, 21, 31, 36, 37, 38, 39, 40, 90, 146
   contracts 39
   towage 37
SAR 190, 214
satellite
   messenger 12, 179, 180
   telephone 1, 9, 10, 178, 180
   tracker 12, 185

sea anchor 31, 88, 92, 108, 118, 121, 122, 123, 175, 183, 190, 201, 203, 206, 208, 209, 216, 248, 249, 262, 263, 264, 265, 267, 268, 269, 270
sea buoys 243
seacock handles 137
seacocks 20, 23, 136, 156, 246
sea foam 230
Seamstik 132, 134, 279
Search and Rescue 178, 214
seasickness 89, 182, 188, 204, 272
securité message 225
sensor arm x, 110
servo-paddle 111
shipkeeping 226
single sideband radio 1, 7
   replacement antenna 11
sludge 52
solar stills 196
solenoid 43, 44, 46, 150, 151, 152, 278
solenoid valves 150
Spanish Reef 97
spreaders 14, 32, 64, 71, 72, 89, 92
squall development 237
SSB distress frequencies 7
Sta-Lok 83
steering iv, 99–124, 272, 277
   chain/cable system 101, 107
   hydraulic fluid leaking 103
   lap zone 106
stern davits 35
storm
   bulletins 234
   sails 251
   tactics 257

stuffing box 44, 46, 278
Supreme Court Computation 39
survival gear 177
swage fittings 66, 81

T

tack viii, x, 74, 76, 78, 97, 164, 165, 166, 250, 251, 253
tangs 64
tank fittings 137
Technora 81, 83, 84, 85
terminal fittings 64
text messages 12
thermostat 53, 57, 58
throttle 20, 44
tide charts 14
tiller arm 103
tool kit xi, 46, 275
tow vs. salvage 35–40
towage 36, 37, 40
towing 21, 24, 25–40, 224
   alongside 32
   astern 33
   bridles 27, 28, 29, 31
      chafing 31
   dinghies 34, 35
   lights 37
   on your hip 32
tow line 25, 26, 27, 28, 29, 30, 31, 32, 33, 34
transmission fluid 43

U

ultraviolet 127
unfamiliar port 13
uninterruptible power supply 4

U.S. Coast Guard contact numbers 9
U-warp 116, 117

V

ventilation 4, 191
VHF battery 11
VHF radio 2, 89, 179, 224, 228
victim retrieval 162, 172
visibility iv, 221–230

W

water pump 43, 53, 55, 56, 57, 145
weather information 2, 235, 241, 243
wheel shaft 101, 107
whisker pole 11, 90, 95, 121, 122, 143
wind vane ix, 108, 110–114, 162, 277
   actuator shaft 114
   servo-paddle 111
   servo-pendulum 111–114

X

Xyalume sticks 211

About the Author

Ed Mapes is a Licensed USCG Master Mariner and founder of **Voyager Ocean Passages** (www.offshorevoyager.com). With 97,000 miles of blue water experience—from offshore yacht deliveries, racing, sail training, and cruising—Ed also provides offshore sailing and cruising seminars at events and major boat shows throughout the country, conducts webcast courses for the Seven Seas Sailing Association, and authors sailing courses for NauticEd. com. As one of only a handful of American Sailing Association instructors certified to train students in the most advanced levels of ASA courses, Celestial Navigation and Offshore Passagemaking, Ed has instructed all levels of ASA courses. He also provides voyage planning consultations to assist those preparing for their own voyages and cruising. Ed has published the books, **Ready to Sail** and **Further Offshore,** and has contributed articles to *Sailing, Sail, Practical Sailor,* and *Latitudes and Attitudes* magazines. Ed has been named a contributing editor of *Sailing* magazine.